NUCLEAR TURNAROUND

Recovery from
Three Mile Island
and the Lessons for
The Future of Nuclear Power

1979-1989

By

William Murray

ISBN: 1-4107-8539-4 (e-book)
ISBN: 1-4107-8538-6 (Paperback)

This book is printed on acid free paper.

1stBooks – rev. 09/22/03

"GPU has come back from the very brink of the corporate graveyard following the 1979 accident. No utility had ever encountered such a catastrophe, either operationally or financially,. so GPU'S management blazed a long and steep trail."

Editor, <u>Electric Light and Power </u>in selecting GPU as Utility of the Year

Throughout this volume General Public Utilities will be referred to as GPU

DEDICATION

This book is dedicated to the employees of GPU and
their families, who during those difficult recovery
years shared in making this a true success story.

TABLE OF CONTENTS

ACKNOWLEDGEMENTS

Appreciation is expressed to the many GPU staff and management members whose interviews provided the personal experiences and opinions reflected in these pages.

W. B. M.

PREFACE

To move from being perceived as the "Great Satan of nuclear power to being recognized for "providing the best customer service in Pennsylvania" is to travel a great distance. It is the distance GPU travelled in the ten year period between the accident at Three Mile Island in March 1979 and its selection in 1988 as Utility of the Year. This document is the record of that recovery.

The accident at Three Mile Island thrust GPU on a lengthy course of corporate and industry "firsts"

- the first utility to experience the consequences of a major nuclear accident.

- the first corporation to face the full impact of the American public's radiation phobia, fueled by media misunderstanding and exaggeration;

- the first utility to overcome the tremendous financial and technical hurdles involved in the cleanup of a destroyed reactor;

- the first utility to restructure
  itself under the new guidelines
  mandated for nuclear power;

- the first utility to overcome the
  many obstacles faced in the restart
  of an undamaged reactor located on an
  accident site.

The pages that follow will discuss the recovery of
the company through the comeback years, focusing on
the areas of nuclear technology, regulation,
communications, financial and legal affairs. It is
not intended to provide a detailed discussion of
the accident itself or every historical event of
this tumultuous period.

However, the reader will note that as the recovery
of GPU, the clean-up of TMI II and restart of TMI I
proceeded online October 9th--- and reached full
power January 86, lessons were learned that were
applied throughout industry. From safety
considerations to regulatory reform, from legal
decisions to communications, what was implemented
at TMI and GPU set precedents across the industry.
The accident indeed has had far-reaching

implications for nuclear power and the utility industry.

The goal of this work is three-fold: First, to provide the reader with some insight into the nature of the company that was able to survive the worst nuclear power accident in the U.S. Second, to put the accident and its aftermath into a perspective that only history and personal experience can provide. And lastly, to offer a look at what the future may hold for GPU and nuclear power now that the TMI chapter is drawing to a close.

KEY RECOVERY EVENTS
1979-1989

The ten-year recovery period was a volatile one, during which GPU experienced a variety of successes as well as setbacks. Listed below are selected critical points in GPU's struggle to regain its position in the utility industry.

April 1979   Unit 2 at TMI safely placed in a safe "shutdown mode," following the March 28th accident.

June 1979   Consortium of 45 Banks agree to provide $412 million to GPU under a Revolving Credit Agreement.

July 1979   GPU presents proposal to Nuclear Regulatory Commission to resume operation of undamaged Unit 1 at TMI, beginning a six year battle.

January 1980   GPU Nuclear created to operate, maintain and manage all GPU System plants.

May 1980	State Regulatory Commissions allow GPU to recover cost of purchased replacement power needed to make up for loss of TMI Units.
June 1980	Venting of Krypton Gas safely conducted, providing necessary first step to allow Unit 2 cleanup to proceed.
July 1980	First manned entry into TMI-2 containment building.
April 1981	First canisters of nuclear waste from TMI-2 shipped to Hanford, Washington disposal site.
July 1981	Governor Thornburgh endorses plan for joint federal, state, and utility funding for the cleanup of Unit 2.
November 1981	GPU announces that its operating company customers saved nearly $400 million through special replacement power purchases made by the company since March 1979.

January 1982   GPU Nuclear licensed by Nuclear
                    Regulatory Commission as operator
                    of GPU nuclear facilities.

February 1982   Bulk of accident-related water
                    removed from TMI-2 containment
                    building.

March 1982   U.S. Department of Energy agrees
                    to accept entire reactor core from
                    TMI-2 for research and disposal at
                    national site.
                    U.S. D.O.E. agrees to
                    provide funds associated
                    with TMI-2 cleanup so as to
                    provide valuable research
                    information.

December 1982   GPU announces that its operating
                    companies have reduced their short
                    term debt to zero from a high of
                    $326 million in 1980.

January 1983   Out-of-court settlement reached in
                    Babcock and Wilcox lawsuit. GPU to
                    receive credit in future services
                    and equipment from B.& W.

April 1983	U.S. Supreme Court rules unanimously that "psychological stress" not be a consideration approval of restart of TMI 1.
May 1983	GPU free of all outstanding bank debt—first time since 1979 accident
November 1983	Admiral Richover and staff complete favorable review of GPU Nuclear and TMI-1 and recommend approval of restart.
November 1983	GPU announces new president and of GPU Nuclear and plans to restructure GPU Nuclear Board of Directors
April 1984	US DOE signs agreement with Japanese providing $18 million for participation in research related to Unit 2 cleanup.
July 1984	TMI-2 reactor vessel head removed giving direct access for defueling.

August 1984   Panel of outside experts say TMI
                   training program "now ranks among
                   top training programs in U.S."

December 1984   US   Electric   Utility   industry
                   announces six year, $150 million
                   program to aid in cleanup at TMI-
                   2. First funds to flow in January
                   1985.

May 1985   NRC votes 4 to 1 to lift the
                   order that shutdown TMI-1 in
                   1979. Opponents appeal NRC
                   action   to   U.S.   Court   of
                   Appeals.

July 1985   NRC report of 6 year health study
                   concludes that 1979 TMI accident
                   had   no   significant   effects
                   warranting further investigation.

October 1985   U.S. Supreme Court votes 8 to 1 to
                   lift stay of operation of TMI-1,
                   allowing   restart   after   6   year
                   delay. Unit 1 goes on line October
                   9th --- and reached full power
                   January 86.

November 1985   Pennsylvania   and   New   Jersey
                 regulatory commissions allow TMI-1
                 back   into   GPU   companies   rate
                 bases. Customer costs reduced $80
                 million yearly across the System.

April 1987   GPU declares first dividend on
                 common   stock   since   cessation   of
                 payments in February 1980.

October 1988   Dividend   tripled   in   1988.   Payout
                 now at same level as prior to 1979
                 accident.

November 1988   De-fueling of Unit 2 scheduled for
                 completion   in   mid-1989.   Plant   to
                 be   placed   in   safe,   monitored
                 storage in 1990.

November 1988 GPU selected as Utility of the Year
        by Electric Light and Power magazine.

CHAPTER I

<u>VIEWS FROM THE TOP</u>

"Following the accident, we slowly and
painfully established a new direction.
Like climbing a mountain, it was not
without its slips and setbacks. But
always pushing on, we finally made it"

William G. Kuhns,
CEO General Public Utilities

<u>A Rosy Outlook</u>

As GPU top management surveyed the future in mid-
March of 1979, the outlook could not have been
brighter. The expensive construction program for
the two TMI plants was winding down and shortly
both plants would be accepted in the rates of both
states served by the GPU System. The fuel mix for
the System's plants was almost equally split
between coal and nuclear, with the expensive oil-

1

fired portion quite small. Customer and shareholder satisfaction were both on the rise. GPU was viewed as a leader in the latest technology for environmental protection and off-peak power savings.

As Chairman Bill Kuhns put it:

> "We had no major concerns.
> We really had it made."

All of this came to an abrupt end in the early hours of March 28, 1979.

A decade later, GPU's chief *officers* could look back on a series of unfolding events never before experienced on the American business scene. This chapter will discuss the recollections of company leaders on the traumatic events GPU experienced in the aftermath of the accident and their thoughts on how and why the company was able to survive and ultimately, to thrive.

An Alligator A Day: Obstacles to Recovery

The days immediately following the accident were riddled with problems that threatened the company's road to recovery, and even its very survival. As GPU managers recall, each day seemed to bring a new obstacle.

The first hurdle, of course, had been the shutdown of the damaged reactor—and it, actually was accomplished safely and effectively. Whatever else happened, GPU management felt secure that the public had not been harmed. In fact, federal and state health authorities have stated that the so-called "worst-accident in the history of nuclear power in the U.S." resulted in no adverse health effects beyond some possible psychological stress.

With the immediate safety issue behind it, the very real practical problems began to pile up: How to continue to run the utility and provide reliable power for two states, with both TMI units out of service? How to proceed through the upcoming legal and regulatory morass to restart Unit 1? How to undertake the massive effort required, financially as well as technically, for the clean-up of Unit II?

These very real issues were clouded by a multitude of largely emotional obstacles thrown up by various groups: the exploitation of communication's problems that damaged the company's credibility in the early days; the accusations and assaults on the company by the very vocal, and often hysterical anti-nuclear movement; gross editorializing and

3

sometimes biased reporting by the media and the inevitable political maneuvering that often follows such crises.

Both the practical issues and the emotional obstacles were first-of-a-kind problems that GPU management faced, and ultimately surmounted.

The Credibility Gap

Early on it was apparent that a major hurdle would be the re-establishment of the publics' trust in the company's words and deeds. The credibility gap grew in the months following the accident as new information became available to the company and was provided to the public. Conditions at the plant became better understood by company engineers. However, when this changing information was communicated to the public, GPU was first perceived as incompetent, then negligent, then so grossly negligent, and finally --- lacking integrity.

Ten years later, trust has not been completely restored and probably never will be, but the high degree to which it has returned is a tribute to the years of dedicated effort by the "people of GPU" over great odds. How far the company has moved from the spring of 1979 when it was considered the Great Satan of utilities to April of 1988 is seen everywhere. The Chairman of the Pennsylvania Public Utility Commission unequivocally declared that GPO's Metropolitan Edison Company was providing the best customer service in the state. About the same time; Wall Street was hailing GPU as the best performing utility stock of the past year. NRC inspections were continually giving GPU's nuclear operation high grades. It has been a dramatic transformation.

The Anti-Nuclear Fight against Restart

Anti-nuclear sentiment immediately following the TMI accident was widespread: the exploitation of "The China Syndrome" movie; the scare headlines threatening "Armageddon"; the entry of the words "meltdown" and "fallout" into our everyday vocabulary.

Much of that sentiment was channeled into a well-orchestrated anti-nuclear effort directed at preventing the re-start of the undamaged unit at Three Mile Island. The ultimate restart of that unit was probably the most important single event in GPU's recovery.

Six and one-half years passed after the accident before TMI-1 was again allowed to generate power. A tremendous struggle ranged from the streets of

Middletown, Pa. to the halls of the U.S. Supreme
Court. There were some real issues and questions
that had to be answered before the approval could
be given, but these paled in comparison to the
thousands of extraneous charges and points raised
by those dedicated to preventing restart. It was
clear from the beginning of this fight that the
anti-nuclear movement would view a go-ahead to
restart a nuclear unit at the site of the TMI
accident as a defeat for their cause. Hence, the
long and tortuous struggle.

Media Coverage and Editorializing

The contribution of the media during these years of
delay must be noted. Those outlets that early on
had dedicated themselves to a negative position on
nuclear power continued to magnify every problem
encountered by the company: minor piping leaks
became full-fledged catastrophes; "what-if"
scenarios became absolute facts in their minds; the
villains running GPU grew in size daily in their
editorial columns.

While professing objectivity, TV news techniques
had the effect of slanting coverage during these
days. Long interviews with company management were
cut to one sentence when put on the air, while
tearful area residents and anti-nuclear scientists

were given full rein for several minutes of
emotional charges. Chairman Kuhns has said:

> "The time and attention the media
> provided to the anti-restart efforts
> were outrageous—giving full meaning to
> the phrase, 'bad news is still the best
> news"

A few examples illustrate the problem: following
the accident there were wild tales of dead birds,
deformed animals, and rustling leaves. Media
treatment of the "hydrogen bubble" episode, even by
the venerable Walter Cronkite, was highly
inflammatory, wrong, and never properly corrected.
Later on, some scientifically inadequate local
cancer surveys were given wide media coverage and
as a result received uncalled for regulatory
attention.

Political Maneuvering

There is little question that this kind of coverage
had a strong effect on the regulators and
politicians involved in considering TMI-1 restart.

Even the Nuclear Regulatory Commission, which has
long been criticized for being in league with the
nuclear industry, seemed to lack the fortitude to
back up its favorable technical findings on TMI and
GPU. The interveners were often effective in

exploiting vacillation and the lack of political strength of the regulators.

National, state, and local politics played a very significant role in GPU's recovery from the accident and the restart of TMI-1. With very few exceptions, GPU faced political opposition at all levels.

Politicians quickly learned that there were votes to be gained if they were publicly identified with opposition to nuclear power and specifically against TMI-1 restart. They knew that their opposition to TMI would gain the support of a very active vocal group—the kind that is the lifeblood of a politician.

Not every politician went along with this however. A notable exception was local Congressman Don Ritter, who put his political career on the line in support of TMI restart. There are other examples of favorable political action but they are few in number.

The TMI-1 restart battle also indicated the amount of power sometimes exercised by the unelected staff of members of Congress—a fact just now beginning to receive national recognition and concern.

Ideologues against nuclear power and with personal biases sometimes used their positions to wield influence far beyond the prescribed limits of their position. In GPU's case, such actions were most evident in efforts to influence the NRC on the issue of the company's integrity on nuclear matters.

Governor Thornburgh of Pennsylvania certainly was a major factor in delaying the restart of TMI-1 with his persistent oral and written opposition. The governor's main interest, which he reiterated time and again, was the cleanup of the damaged Unit 2. He insisted that the cleanup be assured before any thought was given to the restart of Unit 1.

But in the clean-up effort, Governor Thornburgh was of considerable help, lending his name to and strongly endorsing the cleanup funding program. The governor campaigned energetically around the country urging all utilities to participate in the cleanup funding effort.

A Quality Operation

The political battle over restart produced a great deal of anti-TMI rhetoric and inflammatory headlines. The result of all this media hype and political opportunism was years of delay in the

Unit 1 restart approval. In, view of this lengthy struggle, it is interesting to examine the record of that undamaged nuclear plant at TMI, following approval to restart in October, 1985.

In the spring of 1988, TMI-1 was running at 100%. power, providing 860 Megawatts of electricity to thousands of customers. During 1987 the plant produced 75 million megawatt-hours of electricity and operated at a 74.1% capacity factor, compared with an industry average for the previous year of 59.3 %.

An editorial in the Greensburg, Pa. paper on September 8, 1987 summed it up:

> "...It's a quality operation run by quality people...This constructive development has taken place despite all those who wanted to see TMI-1 fail...a failure which could have glared forth as a strong anti-nuke symbol.
>
> ...Most of TMI's neighbors are happy. People are building homes and businesses in the vicinity of the island. Pleasure boaters on the Susquehanna show no fear. Even the anti-nukers and tourists are slowly losing interest in the place."

THE SUCCESS FACTORS

People

Reflecting on the successful turnaround of GPU from its darkest days, the top management of the corporation commented on several contributing factors. It is difficult to single out any one action, event or characteristic that is overriding. But if there is one outstanding contributor to success that consistently appears in the thoughts and comments of the company's leadership throughout these crisis years it is the performance of the people of GPU—from the Board of Directors to those working in the mailroom.

As Chairman Bill Kuhns summed it up:

> "Our people were fantastic. A lesser group would have splintered and run for cover. Somehow we kept our cool, if only because we knew we had integrity and indeed considerably more than our critics."

The remarkable loyalty of the staff at the nuclear plants should be noted. There was no finger pointing of blame so common after accidents of this magnitude. Constructive support towards understanding the accident and solving problems was displayed all up and down the supervisory ladder.

Chairman Kuhns, in reviewing the many trying events of this period, commented:

> "The ability to maintain a sense of humor and an overall perspective under enormous pressures was essential. We were able to sustain a positive attitude and this was reflected in the spirit and morale of all of our people"

During this time the workers of GPU repeatedly showed one overriding purpose—that of providing all of the information possible on the causes and cure of the accident with candor and openness. Considering the interminable interrogations of all concerned and the often unfair pounding of some of the press about integrity and credibility, this alone was an impressive facet of the recovery.

Chairman Kuhns singled out the GPU Board of Directors for special recognition:

> "We were and still are blessed with an extremely. able and supportive Board. As I recall the heavy and numerous problems we tossed at them following March 28, 1979, I am more impressed than ever with their character and competence. The Board performed in a truly exemplary fashion and contributed mightily to our recovery."

Financial Foundation

There is little doubt that the strong financial condition of the company at the time of the accident contributed to its ability to recover. GPU had a good balance sheet. Conservative financial policies had been pursued: the company was not burdened with a lot of leases and had no commercial paper outstanding; no major construction projects were underway, except for Forked River, which was promptly cancelled. The primary financial problem was not new construction commitments, however, but the need to pay for large amounts of purchased power. Devastating as the accident was from a financial standpoint, GPU was strong enough to at least have some time to breathe while the banking and regulatory processes worked.

Innovative Financing

The pulling together of a consortium of 45 banks in June of 1979 to provide $412 million to GPU under a revolving credit agreement stands out as a watershed event in the company's recovery. As elaborated in Chapter IV, these funds provided the working time to survive until relief arrived.

State Regulatory Rulings

Action by the state regulators to GPU's situation often seemed painfully slow, with the threat of bankruptcy continually waiting in the wings. However, when the chips were down the regulators came through and their actions must be listed with the leading factors of recovery. At this time, the regulators utilized as a guiding principle a balancing of the interests of the customers and the shareholders. The understanding and support of the regulators was generated by some outstanding efforts by the GPU staff working this problem. More than once, at the last minute, rate relief saved the day. Here again, the "people" effort mentioned earlier played a decisive role.

Management Decisions

Five specific actions of GPU top management stand out as major contributions to recovery:

- The intensive and lengthy personal efforts by top management in developing and marketing the integrated cleanup funding program.
- The establishment of a separate GPU Nuclear Corporation, with the later

changes in senior management there, naming a new Chairman and President.

- The establishment of a Nuclear Safety and Compliance Committee, comprised of the outside directors of GPU Nuclear, to monitor all nuclear operations.

- The decision to retain Admiral Rickover and his staff to perform an independent evaluation of the readiness and capabilities of the company's TMI-1 operation to restart.

- The application of a policy of complete cooperation and openness in all investigations and hearings.

Of course, the recovery and turnaround of a situation as complex and difficult as that faced by General Public Utilities in the 1979-89 period can be attributed to many, many actions, decisions and events. The following chapters covering technology, financing, communications, regulatory and legal matters provide some expansion of these 'views from the top'.

CHAPTER II

AN UNFORGIVING TECHNOLOGY

"Nuclear power is an unforgiving technology requiring an extreme attention to detail not adequately recognized...It is still a maturing technology with learning from experience being a principal requisite for safety improvement."

Herman Dieckamp President, COO, GPU

It Could Have Happened Anywhere

The accident at Three Mile Island's Unit 2 occurred on March 28 and 29, 1979. It was the result of an extraordinary confluence of events: a minor malfunction in the non-nuclear part of the system triggered several automated responses in the reactor's coolant system. At the same time, a relief valve on the pressurizer was stuck open. A misreading of the symptoms over a two hour period before the relief valve was closed, and the consequent turning off of an automatic emergency cooling system, resulted in the reactor core being partially uncovered and severely damaged.

"It was," as the Rogovin report concluded a year later, "a serious accident." Although there were no deaths, and no injuries have been substantiated, it alarmed the nation largely because, as Rogovin pointed out, "It could have been much worse." And more importantly, "it could have happened anywhere."[1]

Today, a decade later, TMI continues to symbolize for many, rightly or wrongly, the beginning of the decline of nuclear power in this country.

Though certainly serious in itself, the accident at TMI-2 achieved epic proportions because it became more than the event itself; it became the

---

[1] Nuclear Regulatory Commission Special Inquiry Group, <u>Three Mile Island: A Report to the Commissioners and the Public. Volume I</u> (Washington, D.C.; Nuclear Regulatory Commission Special Inquiry Group, 1980) p.5

broad symbol for all potential accidents at any nuclear power plant in the future.

The accident was in fact representative of the state of the industry at the time. Several major investigations at the federal, state, and local levels, notably the Kemeny Report by the President's Commission On the Accident at Three Mile Island, and the Rogovin Report by the NRC's Special Inquiry Group, concluded that the accident involved the entire industrial, technological and regulatory structure of nuclear power in the U.S. The Kemeny Report put it this way: "While the major factor that turned the incident into a serious accident was inappropriate operator action, many factors contributed to the actions of the operators...These shortcomings are attributable to the utility, to the suppliers of equipment and to the federal commission (NRC) that regulates nuclear power."[2]

Shortly after the TMI accident, Admiral Hyman G. Rickover, father of the nuclear navy, commented:

---

[2] President's Commission on the Accident at Three Mile Island, The Accident at Three Mile Island:The Need for Change: The Legacy of TMI (Washington, D.C.:' President's Commission on the Accident at Three Mile Island, 1979)

"Nuclear power will be a better
technology as a result of the lessons to
be learned from this accident."

And the accident at Three Mile Island set in motion
a major reform movement—specifically at TMI-1 and
GPU, and beyond that, throughout the nuclear
industry. (See regulatory chapter.)

The gist of the reforms might be summed up this
way: TMI made people realize that nuclear power was
still a maturing and unforgiving technology. The
changes since TMI involve looking at things we
thought couldn't happen and giving greater
attention to detail in training, planning and
operations, as well as an increased recognition of
the need to learn from experience.

The Event: Coping With A Maturing Technology

Today, the accident can be summarized in a
paragraph. At the time, though, no one understood
precisely what was happening—despite the fact that
GPU, within days of the accident, summoned the top
nuclear experts in the country to TMI to assist.

To begin, both GPU and the NRC were unaware of the
extensive fuel damage. Control board instruments
did not have ability to indicate the degree of
physical damage to the fuel—and so GPU and the NRC

initially assured the world that the minimal radiation releases noted posed no threat to public health and safety.

Two days into the accident, the greater magnitude of the problem became evident. But then, the appraisal pendulum swung in the other direction, and the NRC began to overestimate the situation. On March 30, the NRC announced the existence of a potentially explosive hydrogen bubble in the reactor, which posed the threat of a large-scale release of radioactive material.

Public anxiety—just beginning to ebb after the first announcement of the accident—reached new highs with this NRC news release.

Soon after the initial alarm over the hydrogen bubble, it became clear that there never was the possibility of a hydrogen explosion because the necessary ingredients were absent: the amount of oxygen needed to combine with the hydrogen to make an explosive mixture simply could not be present in the reactor.

GPU, often criticized for withholding information, in fact had no more, and no better, information than what they gave to the NRC and to the public as it became available.

Post-Accident Response—'An Example for the Industry

Over the next several years, GPU led the industry's effort to "mature" the immature technology that had led to the accident. As Phil Clark, President of GPU Nuclear, put it:

> "Because GPU had the accident, the company felt it had to address lessons faster, more broadly, more deeply, better than anyone else."

In the early days after the accident, when the threat of bankruptcy hung over the company like a pall, this was a decision that required a good deal of faith. But, in fact, the leaders of GPU had that faith. Even when it was unclear whether the company could pay its bills the following day, recalls then Treasurer John Graham, those charged with the cleanup of TMI-2 were told by top management to spare no expense—safety came first.

The company underwent major corporate reorganization, advanced the nuclear industry's R & D by years through its cleanup of TMI-2. GPU set precedents for improving the industry's maintenance operations, safety procedures and equipment design with massive changes at TMI-1, leading to its restart. GPU not only wanted to regain the company's credibility, but to become an example for the nuclear industry.

Hank Hukill, who came to the company as Director and Vice-President of Three Mile Island Unit 1 after the accident, recalls:

> "I came out of the Navy nuclear program. I had certain standards that I believed in that this industry had to follow. When I was first interviewed for the GPU position, I asked what the company's ultimate goal was. And I was assured from the top level that their goal was to make our nuclear operation an example in the industry. Nothing would get in their way and they promised me that any resources I needed were guaranteed."

Restructuring

Significant changes were made in both the structure and size of the company to refocus its approach to the nuclear business. The company realized that the nuclear operations must be run by one central organization devoted only to nuclear power. Running a nuclear power plant through a company that operated other kinds of plants was obviously ineffective in meeting the demands of atomic technology.

Hence, the announcement of GPU Nuclear (GPUN), a dramatically different organization, in January, 1980.

Actually, GPU management had already been working on the initial steps in setting up such an across-the-board nuclear plant operating organization in the months immediately preceding the TMI accident. The basic GPU Nuclear organization was established and operating by September 15,1980. Initially it was operated as a "group" under the corporate umbrella at J.G.P.& L, Met-Ed and GPUSC, with its senior officials holding positions in all three companies.

In January 1982, with all regulatory approvals including NRC licenses in hand, it started operating as a regular GPU subsidiary.

As a subsidiary with its own board of directors, the company was charged by GPU with the safe operation of GPU's three nuclear plants at TMI and Oyster Creek. GPUN has full responsibility and authority for all aspects of GPU's nuclear activities and no other responsibilities.

GPU Nuclear's first priority was established as safety. Management at all levels is directly involved in monitoring plant activities. The company is organized to provide inherent safety checks and balances with extensive cross-checking and independent review across disciplines, including oversight by a special Nuclear Safety and Compliance Committee of GPU Nuclear's Board of Directors.

This Committee has its own independent access to plant, operations to look into any and all items of possible safety concern. It has become a model approach for other nuclear utilities across the country.

There are three operating divisions in GPUN—one for each of the plants—and six functional divisions all established to ensure the safe generation of nuclear energy in compliance with all laws, regulations, licenses and other requirements.

People Lessons

Based on the findings of the Kemeny and Rogovin reports, as well as its own internal audit, one of GPU Nuclear's very first charters was to upgrade the professionalism and increase the size of the staff at its plant facilities. AsPhil Clark put it,

> "Nuclear power is very demanding because of the risks...Its fundamental that you must have people who approach it with an attitude that recognizes that risk...By far the most important lessons TMI taught us were people lessons."

New talent was hired to increase the experience level and the total resources, and to add a new perspective to the company's operation. Senior management was chosen on the basis of reputation and veteran experience in the nuclear industry. Many had served under Admiral Rickover in the Navy's nuclear program. Immediately following the accident, TMI management hired only operators with previous experience in nuclear power. Again, about 90 percent had acquired top quality experience through the Navy.

But hiring top quality people was only part of the answer. To fully appreciate the magnitude and importance of the many changes made following the accident, it is necessary to examine a few of these areas in some detail. The following paragraphs spell out some of the specific moves that were made covering people, equipment and policies.

Re-Training

> ".. training is probably the key factor, the key answer"
>
> Hank Hukill

Training was vital in addressing the human element in the running of a nuclear plant. GPU Nuclear recognized a need for greater in-depth training to improve operators understanding of the way a plant runs and improve their capability to cope with the unusual and unexpected.

The Rogovin Report cited most of the shortcomings in training existing prior to the accident. While all operators were licensed by the NRC, written and oral examinations administered to the operators didn't measure an operator's ability accurately, much less ensure that he could operate a reactor safely when something unexpected occurred. There was little on-the.-job training or simulator training.

The simulator training that did take place did not include simulated accidents involving more than a single failure. Nor were operators often required to diagnose the causes of failures given to them on the simulator. The NRC had little staff devoted to monitoring training programs. In short, operator training was a backwater for most utilities and for the NRC. [3]

---

[3] NRC Special Inquiry Group. Three Mile Island, p.103.

The remedy was a sophisticated 10-part training program established at Unit 1. Operators were retrained and relicensed. A separate training division was established.

The training staff increased from seven to over 55 after the accident. A new training facility was built to provide operators with a more professional learning environment. A dedicated simulator costing millions of dollars was erected to provide operators with more hands-on experience. Strong emphasis has been put on the number of hours for training. Each control room crew member spends one out of every six weeks in training. Management involvement is a significant part of training.

Training was improved not only in terms of the number of hours and methods but conceptually as well. A new set of procedural guidelines was set up that focus the operator's attention to a few key symptoms that indicate how the plant is responding to unexpected or abnormal situations. The aim is to have operators respond to the symptoms before trying to identify the specific event, or events, that caused the shutdown.

GPU Nuclear's intensive training efforts have paid off.

As Hank Hukill explains:

> "The reports we've had; not only our own feeling, but the reports we've had from agencies that have monitored us during startup show that the training really paid off. The NRC was here for almost three months around the clock during the start up. They've all noted a remarkable improvement in the knowledge, ability and experience of the operators."

Quality Assurance and Emergency Planning

A greater emphasis has also been put on quality assurance (QA) for nuclear plants. The Kemeny Commission found that "a lack of independent on-site quality assurance or safety assessment of plant operations and of equipment not considered 'safety related' contributed significantly to the accident."[4]

Today, QA is getting center-stage attention from the industry, and GPU Nuclear is a leader in transforming quality assurance from a subordinate of operational staffs into a full partner in the running of nuclear plants. The staffing of its QA department has tripled and its focus has expanded

---

[4] *President's Commission. The Accident at Three Mile Island*, p. 44.

from checking the quality of equipment to monitoring day-to-day operations in the plant.

Poor emergency preparedness involving the public was also identified as a major industry problem in the course of the TMI accident. GPU Nuclear, in collaboration with the Pennsylvania state government and the TMI area county governments, addressed that problem by setting up new and improved plans and facilities for managing emergencies impacting the public. Among the improvements: a full-time emergency preparedness staff; a notification system to the public and government agencies; a siren system for the surrounding areas; revised off-site protection procedures; full-scale drills in Unit 1; a new emergency off-site facility.

The record of GPUN during the many emergency test drills that have been held with the NRC, the state and the federal emergency management organizations following the institution of these programs has been excellent; it has, in fact, stood up against the actions of a very active anti-nuclear group in the area that has continuously scrutinized it for possible lapses.

Equipment Changes

Improved training and expanded safety checks and procedures were a significant part of GPU Nuclear's aim to facilitate interaction between operators and equipment. Upgrading equipment was another. Many modifications have been implemented in the control room and throughout the TMI-1 plant to ensure safety and proper operation of the plant.

Several aspects of the control room have been reorganized. Indicators showing temperatures inside the reactor core have been installed. Three indicators are in place that measure whether the important pressurizer valve that stuck open seconds into the TMI-2 accident is open or not. Control room alarms have been prioritized to help the operators respond to the most important plant symptoms first.

Unit 1's emergency feedwater system, which provides cooling water to the secondary (or non-radioactive) side of the plant if the main feedwater system is not operating, has been modified to initiate automatically if either main feedwater or all four reactor coolant pumps are lost.

Further, an alarm has been installed on the tanks supplying emergency feedwater to give the control room operators ample warning to shift to an alternate source of water. Overall, more than 100 modifications were made. Most of these were required by the NRC as a result of the accident. A majority of the man/machine improvements at Unit 1 and in the entire industry since 1979 are a direct result of lessons learned from the accident.

Preventive Maintenance

Preventive maintenance was also a major part of ensuring safe and proper operation of running a nuclear power plant. Part of the reason for the successful restart of TMI-1 was the constant checking of equipment to ensure its proper performance. The focus at GPU Nuclear is now correctly on preventive rather than corrective maintenance.

Steam Generator Repair

Because of the company's current emphasis on preventive maintenance, there was only one major equipment problem during TMI-1's interim period before restart: corrosion attack on Unit 1's steam generator caused leaking tubes that threatened the plant's future operation.

Every one of the 31,000 tubes had to be repaired individually. The repair job was complicated and became an industry first. Its successful completion represented a landmark for the company and the U.S. utility industry. Phil Clark recalls the event:

> "Essentially all of the tubes had been attacked and leaked. I think one of the things GPUN did which was really outstanding was to devise a way to salvage the steam generators and restore them to essentially as good-as-new condition. This involved in-place repair of every one of the 31,000 tubes by a relatively new process"

Once again, an obstructive anti-nuclear effort was mounted to question the tube repair and prevent approval of TMI-1 restart. Fortunately, the company was able to demonstrate the reliability of its approach to the regulators and this effort failed.

Restart Testing at TMI-1

To ensure that safety measures learned from the TMI-2 accident were correctly in place at TMI-1 before restart, the staff at TMI-1 participated in a costly several-month test program. Hank Hukill recalls why:

"The worst thing that could have happened to us would be to startup and just have problems. Part of the reason we were successful was the test program we used. We had a detailed three month program where we operated at 40 per cent power—that is usual for a plant coming back on line. Forty per cent power for nothing but crew training!

Then we operated at seventy per cent power right thereafter for a month - - - again for nothing but crew training. I think that's a good indication of how the company felt. We were going to prove to ourselves and demonstrate to everyone that we meant what we said. We were going to ensure that we were ready, that people were well trained and ready to operate before we would go to 100 per cent power."

The stellar operating record of TMI-1 since its approval to run has amply demonstrated the correctness of this testing philosophy.

The Cleanup of TMI-2

The cleanup is a billion dollar project. De-fueling of Unit 2 is scheduled for completion in 1989, with the plant placed in safe, secure, monitored storage in 1990. Completion within the planned one billion dollars still appears feasible. The cleanup effort has made, and continues to make, significant contributions to the safe operation of all nuclear plants.

And, the benefits are being sent world-wide since several nations have sent nuclear engineers and technicians to work on the cleanup crews in order to gain first-hand experience in the leading edge technology and methods being employed in the cleanup. Japan has, for instance, contributed 18 million dollars to the project so their nuclear experts could join in the cleanup efforts from start to finish.

The cleanup at TMI involves removal of radioactive fission products dispersed during the accident that contaminated parts of the TMI Unit 2 containment and auxiliary buildings. It also involves the removal of the Unit 2 reactor fuel core, which was destroyed during the accident, and disposal of the fuel and other radioactive wastes.

The radioactive materials being removed from Unit 2 include gases, contaminated water in the containment building basement and in the reactor cooling system, and radioactive particles causing surface contamination. Additionally, there are portions of the damaged core fuel that have been dispersed into the primary system as a result of coolant water circulation that will be removed.

Most gases—primarily krypton gas—were safely vented from the reactor building during the summer of 1980 with negligible impact on the public. A sizable effort was mounted by anti-nuclear groups to prevent this venting, but it was approved when it was shown not to be a health hazard and that it was a necessary step before further cleanup could proceed.

Over a million gallons of contaminated water in the auxiliary building and containment building have been processed and are being stored temporarily on the island.

In 1987, GPUN proposed an evaporative method for disposing of this water. This approach is awaiting final approval. Exposed surfaces in the building areas have been scrubbed down to remove radioactive particles.

An important milestone in the recovery process was the agreement by the U.S. Department of Energy in 1982 to accept the TMI-2 core material for research and development purposes and ultimate disposal at a federal repository.

Although the cleanup has provided important information to the nuclear power industry, it has not resulted in any technological breakthroughs. Rather, the process has been more innovative than revolutionary in that existing technology has been adapted to perform many first-time procedures. As Phil Clark explains:

> "The TMI-2 cleanup is more of a process than a construction...it is a research and development program where you take one step and see what happens and then figure out the next step. Most of it is not routine, but it does not involve technical breakthroughs or major inventions so much as it is a careful understanding of what's there and then applying sound engineering or scientific principles."

The company's response to the special demands of the TMI-2 cleanup has made it an industry leader in the use of robots to solve practical problems. Lessons learned are being passed along to the rest of the utility industry to increase the safety of all nuclear facilities. TMI-2 engineers contribute regularly at industry workshops on these techniques.

An extensive program of manned entries into the Unit 2 containment building to conduct damage assessments, equipment maintenance, radiation surveys and cleanup operations and to explore the most efficient ways of decontaminating affected areas was also an integral part of the cleanup.

Environmental Monitoring

Throughout the cleanup process; the level of radiation releases to the environment and radiation exposure of the TMI-2 staff has been carefully monitored. Test results indicate that cleanup workers have had less collective exposure to radiation at TMI-2 than workers at most operating nuclear plants. Phil Clark assessed the possibility of radiation emission:

> "I think a very important measure of the program's progress is the risk of off-site release of radioactivity as a threat to the public health and safety...There's essentially no risk to the public. And, compared to other nuclear plants TMI-2 is not an on-site threat."

GPU Nuclear has a state-of-the art radiological environmental monitoring program in place around Three Mile Island to detect and assess possible environmental effects from radioactive releases from the plant. In 1985, for example, the company spent almost $1 million to operate and maintain this program run by 20 full-time employees. Since Unit 1's return to service that year, the radioactive liquid and gaseous releases have been a small fraction of regulatory limits. There have been no measurable effects on the quality of the environment from the releases.

Monitored Storage

The cleanup operation at TMI-2 is slowly and successfully winding down. The Monitored Storage status of the plant is expected to begin in 1990. GPU then plans to continue this TMI-2 monitoring until the adjacent TMI-1 is decommissioned At that time, TMI-2 would also be decommissioned. It is expected this will take place well into the next century.

Conclusion:

This chapter began by acknowledging the view held by many that the accident at TMI marked the beginning of the decline of nuclear power in this country. We end by offering a different perspective, suggesting that the outcome of the accident was not altogether negative; that in fact, TMI provided the industry with some plusses. In all likelihood, if the TMI-2 accident had not occurred, there would have been a comparable accident elsewhere. The TMI accident accelerated the necessary maturing of the technology that has taken place in the last few years.

CHAPTER III

REGULATION AND RESTART

"We are convinced that, unless portions
of the industry and its regulatory
agency undergo fundamental changes, they
will over time-totally destroy public
confidence..."

> President's Commission,
> The Accident at Three Mile Island

A Matter of "Mindset"

The President's Commission on, the accident, often
referred to as the Kemeny Commission, named after
the Dartmouth College President who headed it,
remarked that "one word occurred over and over
again" in the testimony they heard on TMI. The
word, though not found in most dictionaries at that
time, was "mindset." Mindset is a mental attitude

so entrenched that it alters reality to satisfy itself.

Kemeny concluded that the mindset that prevailed among people who worked in and regulated nuclear power prior to 1979 was the industry's biggest hindrance. "After years of operation of nuclear plants with no incidents, the belief that nuclear plants were sufficiently safe grew into a conviction."[1]

It is important to realize that this conviction existed among the regulators as well as in the industry itself.

Fred Hafer, currently President of Metropolitan Edison Company put it this way:

> "We were naive, in a sense. The industry was lulled to sleep by such an outstanding performance and safety record."

The nuclear regulatory reforms that were instituted after TMI were designed to change more than regulations: they were designed to alter the

---

[1] President's Commission on the Accident at Three Mile Island, <u>The Accident at Three Mile Island: The Need for Change: The Legacy of TMI</u> (Washington, D.C.: President's Commission on the Accident at Three Mile Island, 1979) p.8

mindset of U.S. industry. The new mindset had to become one which understood nuclear power to be inherently, potentially dangerous. The overriding concern had to become safety. As the Kemeny commission put it: "It is an absorbing concern with safety that will bring about safety—not just the meeting of narrowing prescribed and complex regulations."[2]

Post-accident research studies concluded that the safety systems that existed before the TMI-2 accident put too great an emphasis on equipment and too little on the operators -the human element. The changes instituted after the accident—in training, equipment, control rooms, and plant maintenance— have, as discussed in the previous chapter, emphasized the human element.

In a technical and safety sense, in the construction, operation, and regulation of nuclear power plants, the whole country took a quantum leap forward as a result of the TMI-2 accident.

The structure, purpose and function of those organizations responsible for nuclear power in the U.S. were reviewed closely, overhauled and expanded. Regulators evaluated their approach to

---

[2] President's Commission, p. 9

regulation. In many cases, new agencies and organizations were set up. Functions of existing regulatory agencies were often streamlined and better defined. And while many of the widely publicized changes took place at the federal level, the industry's regulation of itself was probably most profoundly changed. GPU played a major role in the formation, implementation, and response to virtually every one of these changes.

Nuclear regulation, however, still remains under criticism today. Critics charge that some utilities have not learned and applied the lessons of the March 1979 accident. Basic changes in the structure of the Nuclear Regulatory Commission are under consideration. Legislation is being enacted to change the NRC from its collegial format of five Commissioners to a single leader, a structure more consistent with most other regulatory agencies.

The Federal Regulatory Structure

Both the Kemeny and Rogovin reports concluded that the Nuclear Regulatory Commission provided neither leadership nor management of the nation's safety program for commercial nuclear plants. [3]

---

[3] Nuclear Regulatory Commission Special Inquiry Group, <u>Three Mile Island: A Report to the Commissioners and. The Public. Volume I</u> (Washington,

It has been argued that the agency's failure stemmed from the fact that the NRC was concerned more with licensing and promoting nuclear plants rather than monitoring them—and that problem had its roots in the history of nuclear power in the U.S.

The NRC was established by Congress in 1974 for the express purpose of regulating the nuclear industry. It was to replace the Atomic Energy Commission (AEC), because of concern that the AEC's dual responsibility of promoting and regulating the nuclear industry could represent a conflict of interest.

But the "new" agency had precisely the same flaw as the "old" agency—because to a large extent they were one and the same. The same staff remained and the same safety and regulatory processes were adopted by the "new" agency. And, unfortunately, the AEC's approach to regulation became the NRC's. General safety goals were specified and there were prescriptive requirements on design, but their implementation was often left to the nuclear power industry's discretion. There was diverse implementation and a lack of feedback and

D.C.: Nuclear Regulatory Commission Special Inquiry Group, 1980) p. 114

evaluation of the appropriate mode of implementation.

Regulation Undermining Safety

While the attitude of the agency was laissez-faire, there was nevertheless so "voluminous and complex" a mass of regulations that the Kemeny commission concluded that they became "a negative factor in nuclear energy."[4] Besides their complexity, the regulations were too preoccupied with large-break, loss of coolant accidents. Little attention was paid to the human element or to small equipment failures that could slowly build up to a major accident if not handled properly and quickly.

Poor management was another failure at the pre-TMI Nuclear Regulatory Commission. The agency was headed by a fivemember commission, with each member having equal authority. Decisions were reached by majority vote. But, as the Rogovin report pointed out, the members often could not agree, and so the system proved to be a time-consuming and ineffective means of reaching conclusions.[5] In addition, because the offices of the Commission were separate from the staff that ran the agency's

---

[4] President's Commission, p. 9
[5] Nuclear Regulatory Commission Inquiry Group, p.114

daily activities, a strong "we-they" mentality pervaded the organization.

The net result, and the fundamental problem, was that the NRC spent little time deciding broad issues relating to safety. Instead of dealing with policy issues, the commission involved itself in dozens of specific, isolated matters on safety, licensing, personnel and budgetary matters. Thus had come about the maze of regulations that could potentially serve to undermine safety. The effects of this regulatory maze made themselves felt almost daily at GPU during the NRC's consideration of TMI-2 cleanup arrangements and most particularly during the protracted review leading up to the restart decision on Unit 1.

A barrage of safety recommendations issued forth from official investigations, and government officials often felt compelled to set arbitrary deadlines for action. Many suggested improvements proved to be less than cost-effective, often impractical, and sometimes unnecessary. It almost seemed that the NRC changed its regulations daily. They would superimpose the new regulations on plants that were in service and on plants that were under construction. The confusion level and the cost impact were very high. Some post-TMI rules became safety concerns in themselves because they imposed requirements and deadlines for implementation without a realistic analysis of their true contribution to safety.[6]

Restructuring of the NRC

Post TMI, the NRC underwent considerable change, strengthening its regulations and requirements, increasing the number of inspectors assigned to nuclear plant sites and increasing the financial penalties it charged for alleged safety infractions.[7]

Since 1979, the number of NRC inspectors in the field has increased substantially: over one hundred NRC resident inspectors work at 79 sites around the country, compared to only a handful of investigators before TMI. And in June 1980, an amendment to the Atomic Energy Act gave the NRC the power to impose fines as high as $100,000 for each day that a nuclear utility fails to meet federal requirements governing nuclear power operations. The intent was to make compliance cheaper than non-

---

[6] Atomic Industrial Forum, Inc., <u>The New Reality: Update on Nuclear Plant Safety</u> (Bethesda, MD, 1983) p.20
[7] IBID, p. 5

compliance. Previously, civil penalties were limited to $5000 a day and $25,000 a month.[8]

In general, post TMI technical improvements mandated by the NRC call for work to be conducted in stages - - for example, a design review or operations analysis comes first, followed by proposed modifications, and lastly, the actual changes are made.[9] Though this ensures greater safety, the general public, or the critic eagerly looking for flaws, this careful procedure is sometimes unfortunately viewed as foot-dragging.

On the whole, the industry has done remarkably well in satisfying all of the new requirements. The immediate corrections that were made at TMI were paralleled at most nuclear installations. Of the 130 specific safety fixes called for in the government's roadmap of post -TMI nuclear plant safety, more than 70 percent were made by 1987. The remainder, mostly long range improvements, are under government and industry review.[10]

Though a few in the industry still believe that the NRC regulations are too cumbersome, the past-TMI trend of stronger regulation is likely to continue.

---

[8] IBID, p. 15
[9] IBID, p. 20
[10] IBID, p. 6

The solution, as Bill Verrochi, the retired president of Jersey Central Power and Light (owner of the Oyster Creek Nuclear Plant) pointed out, is cooperation:

> "We, the people in the utility industry, have to learn how to work more effectively with the bureaucrats, the staffs and their leaders. Learn how to make the process work. Recognize that it is going to take a lot of time to get things done. Plan far enough in advance so that the inevitable delays don't hurt too much."

For the most part, those failures by nuclear utilities In the years since TMI can be traced to a lack of full recognition of the "human element" in the implementation Of regulations, rather than failures in plant and equipment.

Wrestling with Restart

The six year plus struggle to obtain approval to restart the undamaged TMI Unit 1 was testimony to the fact that the structure of nuclear power regulation in the U.S. is in need of further streamlining.

And although a multitude of factors contributed to the delay of restart - - including opposition by special interestgroups and publicity opportunists, numerous lawsuits, and technical overhauls, the regulatory morass was a significant contributor. That morass was thickened not only by the complexities generated by the NRC's regulatory process itself but also by its ambivalent posture on restart. There often seemed to be a lack of conviction to move ahead on conclusions once reached. The NRC had recommended restart fairly early on - in December of 1983; and two years earlier, the NRC's licensing Board, the ASLB, had also sanctioned restart.

Certainly, in all fairness, the NRC had very real issues to contend with during this period. Among them were problems that either arose or remained unresolved after the initial TMI restart recommendations. Primary among these were the falsification of leak rate data at TMI-2 before the accident exam cheating episode. (Both of these are discussed further in Chapter V.) Under any regulatory structure, were issues that would take considerable time to sort out and either disprove or correct.

Additionally, there was one nagging open issue that seemed to run on and on. This was what the NRC called the question of the integrity of GPU management. The integrity concern reached a low point for the company during April and May of 1983 when the NRC informed the company that, despite months of prior investigation and hundreds of hours of on-site inspections by their staff, they could not re-validate their earlier favorable finding and thus could not reach a conclusion on the issue of management integrity and competence.

Two years and five months would pass before Unit 1 at TMI would finally receive the go-ahead for startup. Major changes by GPU in its management structure, appointment by GPU of an independent safety committee, countless more inspections, days of hearings evaluations by several internal and external specialists were required before these issues were put to rest.

The NRC's regulatory process in itself imposed many formalities which greatly lengthened and undermined its ability to reach conclusive and timely decisions relating to restart. One of the great artificialities of the restart process was that there was an ongoing formal proceeding which addressed licensing, management, and competence but which afforded little opportunity for the decision makers to directly judge the integrity of the company. In other words, the NRC process, during pre-restart, was simply a public tribunal and did not encourage cooperative efforts at understanding and problem solving. And finally, the five commissioner structure served to undermine the agency's focus on the real issues. For example, when NRC Commissioner Gilinsky issued his personal statement in June of 1983, citing what he maintained were shortcomings of GPU and demanding the dismissal of GPU's top management, a good deal of time, energy, and attention at both GPU and the NRC (not to mention the press) was diverted to what was essentially a personal matter made public. Ultimately, Commissioner Gilinsky's term on the NRC expired before the NRC vote on the restart of Unit 1.

Once the NRC approved restart, it went to great lengths to ensure its success, nearly doubling its monitoring standards both before the plant went into operation and after restart.

As Hank Hukill described it:

> "The normal plant gets an NRC inspection once a year. We've had one and we're getting another one. We were monitored around the clock for three months. They've reduced that from 24 to 16 hours a day. Most units have one, maybe two resident inspectors on site. We have four."

And the net result, Hukill continues, has been positive for GPU because it has made the NRC understand "where we're coming from. That all of GPU is supporting the safe operation of this plant."

Industry Action

TMI brought the industry together for action. Within hours after the accident, nuclear industry experts across the country were contacted by GPU technical management. Initially, their expertise was brought to bear on the immediate problem of safely cooling down the damaged core. Many of the nuclear utilities and the national laboratories quickly sent their brightest and best to TMI to help out.

This task force of experts was set up outside of Harrisburg in the buildings of the Pennsylvania National Guard near the TMI site. They worked around the clock for the next few weeks. Their accomplishments are one of the several unsung, almost heroic, efforts of those difficult days. That group and its actions were indicative of the new spirit of cooperation and joint support that was to follow amongst the nuclear utilities and the manufacturers in the months and years ahead.

Two policy committees composed of top utility executives were soon established: The TMI Ad Hoc Nuclear Oversight Committee and the Atomic Industrial Forum's Policy Committee on Follow-up to Three Mile Island. This was the beginning of a decade of industry collaboration to improve the safety of nuclear power.

The Long Term

A new feeling had now developed among nuclear organizations nationwide. A problem at one plant was now recognized as a matter for concern at all plants - - not just from a technical standpoint, but from a political and a public relations point of view as well. As Bill Moyers was to express it on the CBS Evening News almost seven years after the TMI accident, "...We all live on Three Mile Island.." Chernobyl, of course, reinforced this "one world" approach to nuclear safety.

Three important institutions marked the nuclear industry's collaborative efforts toward ensuring safety: The Nuclear Safety Analysis Center (NSAC), the Institute of Nuclear Power Operations (INPO) and the Nuclear Utility Management Resources Committee (NUMARC)

Each group deals with the technical or operational aspects of nuclear safety, yet each was designed with a specific goal in mind. NSAC was formed to study the accident and nuclear plant safety questions, while INPO was established to promote excellence in the management and operation of nuclear plants. NUMARC was set up to facilitate communications within the industry and with the NRC.

Overall, these organizations have served to accelerate safety analysis and planning and to create an environment of safety consciousness in the industry. Their importance and influence has grown with each passing year. An unfavorable INPO nuclear plant inspection report can have a telling effect on utility management.

The GPU Role

Senior GPU nuclear managers and staff members brought to bear their solid backgrounds and experience in the activities of all these organizations. Their contributions to the success of the industry groups, to GPU's own recovery, and to a growing rise in public confidence can hardly be overestimated.

GPU staff were naturally staunch supporters of any recommendation or modification that could enhance nuclear safety. But equally important, they spoke out against hardware or systems that, though recommended in the name of safety, in reality contributed nothing or, even worse, could detract from real safety.

From the very beginning, GPU cooperated actively with the goals and operating programs of INPO. Their plant inspection and grading teams were given total company support and assistance as they completed their independent audit evaluations at TMI and Oyster Creek.

Recognition of the organizational, personnel, and procedural strengthening at the GPU nuclear sites was noted favorably time and again by INPO inspectors. A searching analysis in 1981 by an INPO audit team found the TMI-1 plant ready for safe operation. GPU was complimented for its commitment "to acquire top-level talent for its operator training, quality control and communications."

In the important area of training, GPU stressed obtaining formal INPO accreditation (a process for nuclear utilities similar to that used for colleges and universities) for its operator training programs under INPO's strict evaluation program. In 1984, a separate panel of outside experts reported that GPU Nuclear's operator training program "now ranks among the top utility programs in the United States

Government/Industry Cooperation

The government and utility groups that were set up as a result of the TMI accident work closely together. It is to each party's advantage to streamline activities in order to instigate and implement the most effective and efficient safety improvements, to pool together talent and money from both sides in order to optimize all resources and to capitalize on the R & D opportunities that the accident cleanup provides.

LEND is an acronym that collectively identifies the four utility and government organizations that are cooperating in the TMI-2 cleanup: GPU; EPRI (the research arm of the utility industry), the NRC, and DOE (the federal energy research and development agency). LEND is a good example of how the-collaborative industry/government team works to achieve the ultimate goal of bettering nuclear power technology.

Frank Standerfer, Vice President of TMI-2, describes the government/industry cooperation in the NRC and DOE relationship with GPU in its accident cleanup program:

> "The NRC set up machinery to allow us to plan, recommend, and get approval for cleanup work that had to be done. We're still under a license, and in that sense we have to get their agreement that what we're planning to do is safe. Yet the normal way of doing that under a license couldn't be used here.

The Department of Energy, as the other partner, is providing funds for a lot of the work. They are taking a lot of the waste that we can't. The utilities through their research organizations are also funding some cleanup activities."

Commenting on another cooperative effort underway, GPU Nuclear President Phil Clark considers NUMARC, the Nuclear Utility Management and Resource Committee," one of the most important things that has happened to the nuclear utility industry in years." NUMARC is comprised of senior executives from the industry's 55 nuclear plants. Clark plays an active role in this group's work.

NUMARC's original purpose was to assist in shifting primary attention from design approval for plants under construction to focusing on plant operations.

One of NUMARC's initial tasks was to request the NRC to support industry efforts in management and people-related areas and to issue appropriate policy guidance rather than unnecessary detailed regulations. The NRC agreed. This marked a turning point in the search to avoid overregulation.[11]

The Rickover Reports

GPU did not limit its safety reviews and evaluations to cooperation with the NRC and the industry-sponsored organizations outlined above. Several strong internal actions were also taken to ensure that additional independent evaluations of safety performance were made.

In 1983, GPU Chairman William Kuhns commissioned Admiral Hyman G. Rickover and his staff to make an independent assessment of the GPU Nuclear organization, including the soundness of its organization and senior management, in anticipation of the return to service of TMI-1. In November of 1983, Admiral Rickover issued his first report, concluding that GPU Nuclear Corporation "has the management, competence, and integrity to safely

[11] American Society of Nuclear News, January 1986

operate the TMI-1 plant." A re-inspection by Admiral Rickover and his team the following year confirmed his favorable recommendations and again gave TMI-1 overall high marks.

Safety Committee Formed

In 1984 three outside directors were named to the GPU Nuclear Board. These three men, each one a nationally recognized expert in his area of nuclear power, comprised a Nuclear Safety and Compliance Committee. The Committee was provided with its own independent staff to perform evaluations and make recommendations directly to the GPUN Board of Directors. Their reports are also made available to the NRC.

On October 9th, 1985 the long and tortuous review and regulatory process finally reached its successful conclusion with the restart of Unit 1 at Three Mile Island.

Conclusions

While the foregoing review describes several of the post-accident changes in nuclear regulation, particularly as they relate to TMI-2 cleanup and TMI-1 restart, the field of nuclear power regulation requires more to be done on a national basis.

In 1986, for example., the NRC was cited on several counts in an audit conducted by the U.S. House Committee on Appropriations. The report claims that virtually none of the reforms suggested by the Rogovin and Kemeny Commissions for the NRC have been implemented, and says the NRC's post-TMI reform effort has failed. The industry strongly disputed these conclusions and felt the report was extreme. Nevertheless, incident reports at nuclear plants continue to show that the TMI lessons have not been completely absorbed.

Other areas of regulation and review call out for change. The corporate carnage that has been wrought as a result of the handling of the Seabrook and Shoreham nuclear plants in the Northeast points up the need for review of the methods for handling the approval of plant evacuation planning in the licensing process.

Most importantly, the reforms in government and industry that began after the TMI accident must serve ultimately to restore public confidence in nuclear power - - if nuclear power is to survive as an energy source in the U.S.

CHAPTER IV

FIGHTING THE FINANCIAL FALLOUT

"If someone had sat us down right after
the accident and told us what was in
store for us...we might have concluded
that the pressures would be too
great...that all the things that would
have to be done would just be
insurmountable..."

Fred Hafer President, Met-Ed

Uncharted Waters

At the beginning of 1979, GPU was in better
financial shape than it had been in ten years. The
stock was selling at $18 and paying a dividend of
10 per cent. Construction of the two units at Three
Mile Island, totaling 1770 megawatts, was complete

after ten years of work, and represented one fourth of GPU's capitalization.

Those units also represented in a significant contribution to earnings: in January, both Jersey Central and Penelec had received rate orders reflecting in their base rates the capital and operating costs associated with TMI; Met Ed, the owner of fifty percent of TMI-2, received rate order for that unit in early March and expected to begin collecting the last week in March. With things in this kind of order, GPU had reduced its line of credit from $250 million to $225 million, and told its banks it would be at $200 million by June.

The only financial concern of the day was the need to increase equity, which was thin; even that was an industry-wide issue that all utilities were only beginning to address.

While there has been considerable discussion as to the degree of melting that took place in the reactor core during the TMI accident, there is no doubt that the financial structure of GPU melted rapidly in the days following the March 1979 accident. The accident undid it all. But more than that, it precipitated financial issues that had never before been dealt with by a public utility company.

The Core Problems and The Critical Solutions

The characteristic feature of the financial situation created by TMI was that the problems tended to compound one another, and the solutions tended to be bundled, like a tangled skein, so that survival depended on a multitude of factors coming together at precisely the same time.

John Graham, then GPU Treasurer, expressed the situation this way:

> "We could have withstood the nuclear accident; that was, relatively speaking, 'a piece of cake'...But when you compounded that with the price of oil at the time—about $30 a barrel, and inflation going crazy...and then compounded that with Oyster Creek, causing a very serious demand for capital on the one hand and pushing up Jersey Central's rates on the other...it's the compounding of all those that was the real problem."

The nature of being a public utility and of being licensed to operate a nuclear facility imposed special, demanding financial obligations on GPU. A delicate balancing act was required. The company had to balance two competing demands in the face of very inadequate cash resources. First, it had to continue to maintain service, which meant buying energy to supply customers at rates considerably higher than the customers were paying: the bill for replacement power totaled, $24 million a month, at the same time that both Units 1 and 2 were removed from the rate bases. And second, it had to protect the public by stabilizing the damaged reactor and beginning the decontamination. Most businesses would have had the opportunity to reduce their sales when their low-cost supply was cut off; a public utility could not cut back on providing electricity. Most businesses could have postponed rehabilitory work till funds were available; GPU had to pursue the cleanup at TMI 2 at any cost.

So as expenses rose sharply, and income fell sharply, GPU was largely dependent for its survival on the decisions ofthe state utility commissions, which not only directly impacted the company, but indirectly influenced the decisions made by the bankers, another group critical to GPU's financial survival.

Financial survival depended on restoring TMI-I to service and to the rate base. This would maintain GPU's required objectives of providing reliable, economic electrical service to the region <u>and</u> assuring that shareholders did not suffer unduly.

The unraveling of the tangled skein to achieve its goals is, to a large extent, the record of GPU's survival.

The Balancing Act

Though the various solutions GPU pursued were in and of themselves fairly straightforward and immediate, in most cases the success of one was contingent on the success of another.

On the weekend after the accident, for example, GPU contacted its banks, asking for an increase in the informal lines of credit from $225 million to almost $500 million.

The immediate reaction of the bankers was tremendously supportive. As John Graham, who at that time had just begun his position as treasurer, recalls:

> "There was virtually a unanimous indication that our lines would be increased and that the banks would stay with us. I believe that reaction of our banks to be a compliment to the American banking system."

Yet as the situation evolved, and GPU needed to establish a long-term revolving credit arrangement with 40 banks, the bankers began to be very sensitive to rate decisions.

Another important step in surviving the immediate impact of the accident was the securing of permanent financings—which given the company's situation at the time was no small matter. No market existed for Met Ed's securities, but GPU was able to place $50 million of Jersey Central bonds and $50 million of Penelec bonds. But here again the skein began to tangle. The placement of bonds became contingent on two conditions: first, putting in place the banks' revolving credit agreement, which was still under negotiation, and second, rate relief action from the state regulatory agencies that would ensure Penelec's and Jersey Central's continued viability.

John Graham recalls the tenseness of the situation:

> "We found ourselves coming down to the third week in June when a very substantial PJM Interchange bill for power purchases would be due and we would not have the cash resources to pay that bill.
>
> However, we would get bank credit to pay the bills if we got rate orders from two

commissions deemed adequate by a group of more than 40 banks (each with different interests and priorities). We would then be able to close on $100 million of permanent financings—if we could get that bank credit and those rate orders.

Both commissions had been scheduled to make decisions on June 15th. Then one of the commissions announced that it would not make its decision until the other commission made its decision.
I think you can understand that there was a level of tension within the GPU organization by that time."

One of the ironies of timing in the bond sale was the fact that the final sale was made the same week that the all-important report by the President's Commission on the TMI accident (the Kemeny Report) was released.

The timing of these various requirements and events was so fast that often it was simply impossible to adequately communicate throughout the organization. GPU found that it had some of its people out talking to bankers, regulators and investors with information that was correct the day before but which had become incorrect overnight.

The Darkest Days

During these post-accident crisis days, the state regulator's decisions were surrounded by a tremendous amount of emotionalism—particularly in Pennsylvania. From almost the day of the accident, GPU was threatened by calls for punishment and retribution. There was a cry for scalps.

Amidst this emotional environment, GPU was dealt a nearly lethal blow by the regulators. The capital costs for both TMI-1 and TMI-2 were removed from the rate bases. Initially, because it was assumed Unit 1 would be allowed back in service by 1980, this decision seemed less severe than it turned out to be. The decision was appealed to the Supreme Courts in both New Jersey and Pennsylvania, but to no avail.

As restart began to drag, however, the omission of Unit 1 from the rate base became critical to CPU's survival. A regulatory decision to allow amortization of Unit 2, however, was an important contribution to the company's cash needs.

Emotionalism, and the need for punishment, took several forms beyond the expected one that the company should pay all costs for the accident. Met-Ed was ordered to show cause why it should not lose its franchise to operate. "Fault" investigations that were to drag on for several years were started. And there was a call for immediate, comprehensive management and financial audits.

In Pennsylvania, the PUC ordered Theodore Barry Associates to conduct a full scale audit of Met-Ed, Penelec, and GPU, and to reach some quick conclusions on the financial survivability of Met-Ed and GPU. In New Jersey, the Board of Public Utilities (BPU), concerned about the long-term capability of JCP&L to serve its customers, brought the firm of Arthur Young on board to look into various organizational alternatives for GPU's New Jersey company. These early days were also full of merger rumors and a possible state power authority takeover.

Touch and Go Days

The early days also held the threat of bankruptcy - a possibility which the media picked up and would not let go.

At one point, when the revolving credit agreement was pending, as were the rate orders, as were the bills for the replacement energy, there was one weekend of very hectic negotiation for a $40 million, 60 day loan that Jersey Central needed to pay its May power interchange bill. While that negotiation was going on, Bill Murray, GPU, Communications VP recalls:

> "We had two press releases drafted. One said GPU had been able to pay its power bill. The other said the company had gone into default because it did not have sufficient funds available to pay its bills."

The worst fear, according to John Graham, was that by discussing the possibility of bankruptcy, we might create a self-fulfilling prophecy. He adds:

> "It almost happened. Bankers and suppliers-but particularly bankers - saw the publicity about bankruptcy and, as a result, moved to cut off or restrict credit to the system.
>
> Had we not told the regulators about the reality of our limited access to credit, we might well have had more difficulty in obtaining the rate orders. It was a 'dammed if you do and dammed if you dont' situation. I am convinced we took the right course of action—but I know there are many who believe we were just trying to stampede the regulatory commissions and others who think we overreacted and unnecessarily created turmoil."

Outside Studies

During 1980 and '81 several outside studies were made of GPU's financial situation. A study by the Arthur Young and Company consulting firm concluded that bankruptcy was not a desirable solution to the company's woes. And in 1981, the U.S. General Accounting Office completed the second of two financial studies[1] confirming that conclusion, and also recommending that the Department of Energy come up with a long range, federal research and development program to assist in the cleanup of TMI-2.

The Beginning of Recovery

The turmoil began to subside in 1981.

First, the company had come forward with its own plan for reorganization in early 1980. A separate GPU Nuclear Corporation was set up with sole responsibility for operating GPU's nuclear plants. Ownership of the plants would remain with the three GPU subsidiary companies. The plan went ahead. GPUN was formally established after state and NRC

---

[1] Reports by the Comptroller General of the U.S.: Three Mile Island: The Financial Fallout EMD 80-89 June, 1980 Greater Commitment Needed to Solve Continuing Problems at Three Mile Island EMD 81-106 Aug 1981

regulatory approval was finally received two years later in 1982. Nothing had come easily.

The GPUN organizational pattern for the operation of nuclear power plants became the model for several other troubled U.S. utilities in the years ahead.

No doubt the critical junctures were the decisions by the state regulators that permitted GPU to survive: allowing GPU to recover the costs of replacement energy and allowing the amortization of TMI 2. The latter was an important decision for the entire industry because it said, as Fred Hafer explains:

> "...if you spent money prudently, even though things didn't work out, you should be allowed to recover your investment."

It was also, conceptually, a reinforcement of GPU's position that the costs of the accident should be shared by all parties: the company, its shareholders, the customers and the government.

It is interesting to note that now, ten years later, we find several utilities involved in lengthy "prudency" reviews to justify their early investments and spending for nuclear plants—usually because of a perceived near-term energy excess. It seems there will always be Monday morning quarterbacking.

Sharing Cleanup Costs

In July of 1982, Governor Thornburgh of Pennsylvania endorsed a cost-sharing plan for the decontamination of TMI-2. This plan, which bore the Governor's name from that time on, looked to the participation of both national and local sources for the funding of the decontamination effort.

Because of the research and development aspects of the effort, half of the money would come from the federal government and the utility industry. Because of the interest of the involved states in the elimination of the health and safety hazards at the Island, a combination of the company's ratepayers, available insurance, and the two states would contribute the required other half. In addition, the federal government committed to taking possession of the core fuel once it was removed from the reactor vessel and storing it in a national facility.

International interest in the valuable lessons to be learned from the TMI cleanup became evident in 1984 when it was announced that seventeen Japanese electric utilities would contribute $18 million over a five year period to participate in the USDOE research and development effort on the TMI-2 reactor.

Internal Cost Cutting

Survival also required internal cuts, and resulted in some casualties. Bill Verocchi, then president of Penelec, recalls the task of reducing internal spending that required "cutting the heart out of construction expenditures and out of operating and maintenance expenditures." The ultimate casualty, of course, was cancellation of the planned Forked River nuclear project.

Cutting the GPU dividend was painful. Reduced in 1979, the dividend was omitted—for the first time in GPU history -beginning in the first quarter of 1980. In light of the enormous financial burdens on the company, and following the concept of shared cost, it was a decision that had to be made. However, it raised a larger issue of future investment in a regulated industry. As John Graham commented:

> "In setting a rate of return in the range of 13% or so, and in setting rates that have allowed earned returns of 10%. or 11%.., I do not believe the

74

regulators ever indicated to investors that they were bearing this kind of risk. If continued into the future, this decision will seriously and adversely burden energy in New Jersey and Pennsylvania and, I believe, nationwide. The capital will simply not be available to the industry at the levels of return that have been allowed if this kind of risk might fall upon a limited group of investors."

And it led to a class action suit by shareholders (see next chapter), and to some of the most well-attended and raucous annual meetings in the history of the utility industry. By and large, though, the shareholders of GPU remained supportive of the company.

The Cogeneration Contribution

Matching energy demands and capacity in a cost effective manner during this turnaround period was a persistent challenge. GPU continued to implement its master plan for energy supply first developed in 1969. Cogeneration was a key element of this plan. (Cogeneration is the use of previously wasted heat from non-utility industrial processes to generate electricity to be fed into utility lines.) Combining this program with a major conservation effort and off-peak rate incentives, GPU was able to head off the need to add 750 megawatts of generating power to the system, almost the equivalent of building a nuclear plant the size of TMI-1.

At the 1988 annual meeting President Herman Dieckamp predicted:

> "GPU's dependence on cogeneration and independent power production is expected to increase significantly in the future, accounting for about 28 % of the system's requirements in 1997."

Turnaround Painfully Achieved

The worst of the financial crisis was over by the mid -80s, GPU's operational and financial health was giving all of its publics (customers, shareholders, regulators and legislators) confidence in its ultimate return to whatever passes for normalcy in the utility industry these days.

The state regulators, although often exasperatingly slow, continued to break new regulatory ground in their measured support.

By May of 1983, GPU could announce that it was free of all outstanding bank debt for the first time since the TMI accident.

Persuaded by extensive personal efforts from GPU top management, the electric utility industry across the country was chipping in financial support for the TMI-2 cleanup effort. The return to service of Unit 1 in 1985 was making a major dollar contribution to financial operations. Internal cost cutting continued, combined with favorable power purchases from other utility systems. All of this added up to an increasingly positive financial climate for General Public Utilities.

In April 1987, the company announced the restoration of a dividend for its shareholders, albeit a small one. In less than a year that dividend was doubled, and at the 1988 annual meeting, Chairman Kuhns was able to forecast continuing dividend increases. The common stock was selling at over six times its price in 1980, making it the best performing utility stock in the nation.

The industry press has been lavish in its praise of this turnaround. As the editor of <u>Electric Light and Power</u> put it:

> "GPU has demonstrated that a nuclear
> utility can survive a potential reactor
> disaster, weather the many storms that
> follow in the wake of total disclosure
> of its weaknesses and its sins of both
> omission and commission and still serve

its publics without pause. It has shown
the dedication of its people at all
levels, underlining its commitments to
customers, regulators, investors and the
environment. In this, GPU has performed
a service of inestimable value for the
entire industry and for the general
public throughout the nation."[2]

What probably was read with even greater
satisfaction by all GPU employees were the final
words of a local Pennsylvania newspaper editorial
in May of 1988:

"It is truly heartening to report that
GPU and Met-Ed have risen from the ashes
of TMI to regain the high regard they
once commanded."[3]

---

[2] Electric Light and Power; Editorial, August 1987
[3] A.M. BERKS;May 7,1988

LOCATION

Three Mile Island, in the Susquehanna River, Londonderry
Township, Dauphin County, about 10 miles south of Harrisburg,
Pennsylvania.

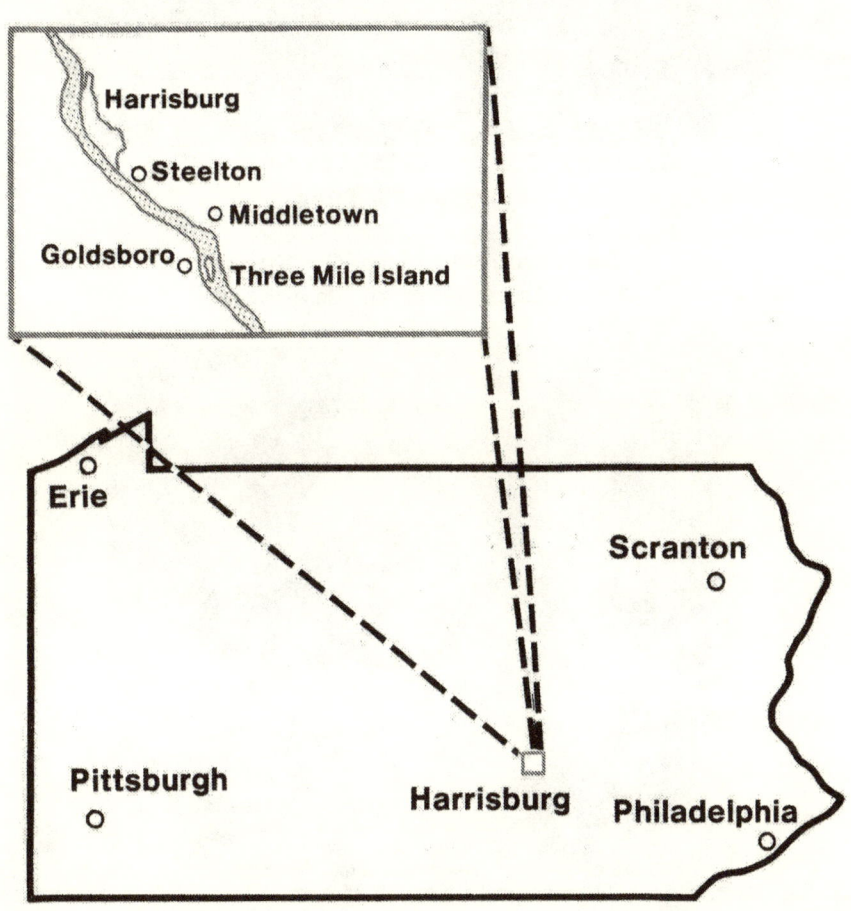

79

General Public Utilities (GPU)
Management Team

Chairman William G. Kuhns
President Herman Dieckamp

John Herbien, Metropolitan Edison V. P. Generation
Early accident spokesman for GPU

Members of the Presidents Commission on the TMI
Accident tour Unit 2 Control Room

On the left, Commission Chariman
John G. Kemeny, President
Dartmouth College

President Jimmy Carter and Pennsylvania Governor
Dick Thorhburgh entering Unit 2 Control Room during
the TMI accident

William Murray

**TMI Supervisors and Control Room Operators Confer
in Unit 2**

TMI-2 CLEANUP OPERATIONS

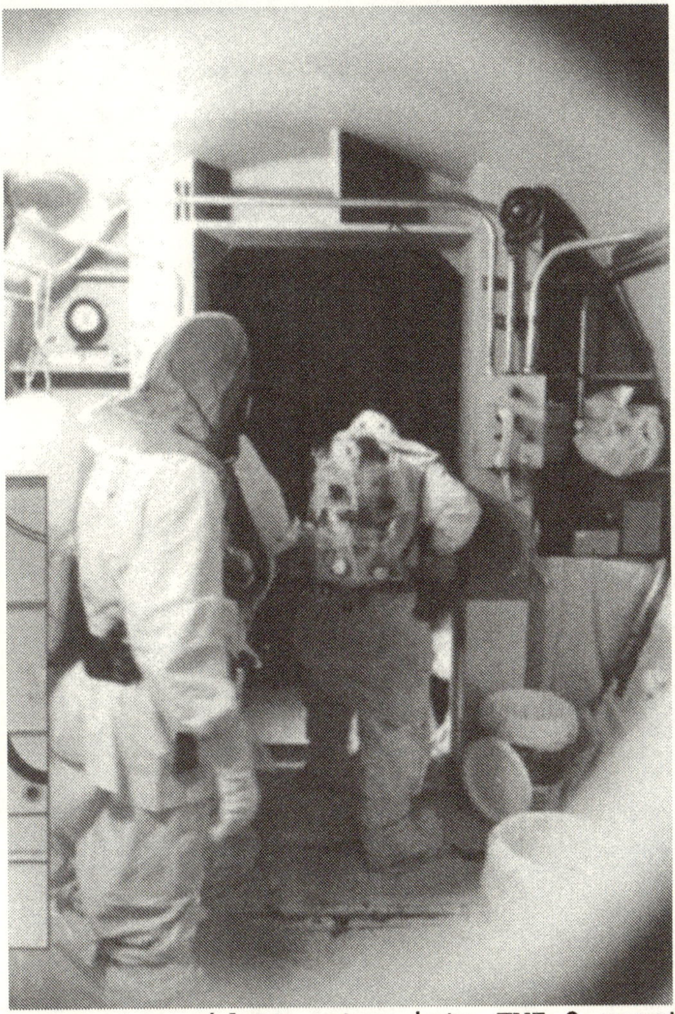

First post-accident entry into TMI-2 reactor building

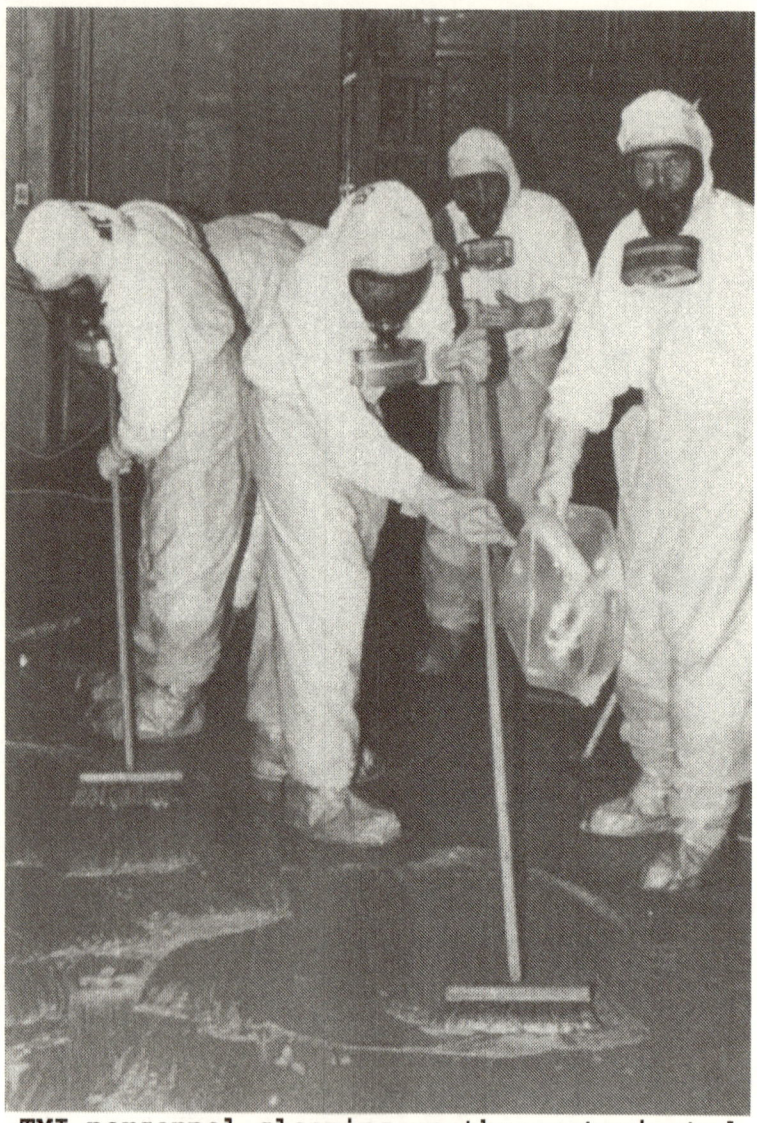

TMI personnel cleaning up the contaminated
auxiliary building

RESTART OF UNDAMAGED TIM UNIT 1

In November 1983, GPU Chairman Kuhns requested Admiral H. G. Rickover and associates to independently assess the soundness of the GPU organization and its management and its competence to operate the undamaged nuclear plant at Three Mile Island (TMI-1)

Admiral H. G. Rickover
Chief, U. S. Navy Reactors Branch

Rickover concludes that the GPU Nuclear Corporation has the management, competence and integrity to safely operate TMI-1...the plant should be permitted to start operation.

TMI-1 resumed operation October, 1985

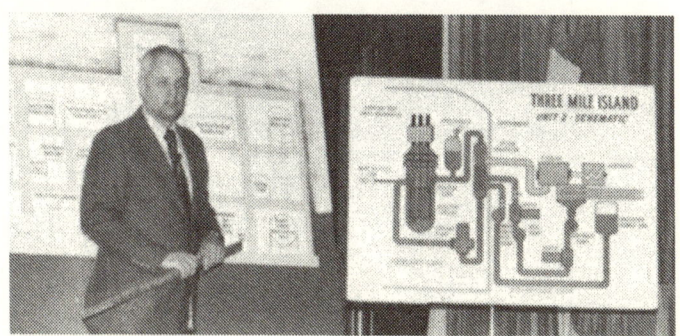

R. C. Arnold, head of the TMI Restart and Recovery Effort

Phil Clark, President GPU Nuclear Corporation

Walter Creitz
President, Met-Ed
(during accident period)

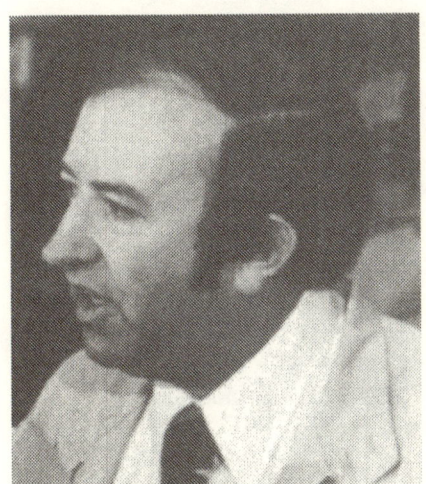

Harold Denton
Director of NRC
Office of Nuclear Reactor Regulation
(primary spokesman for NRC)

Fred Hafer, President
Met-Ed (post accident)

John Grahan, GPU Corp VP
Treasurer

Bill Murray, GPU Corp VP
Communications

ADVERTISERS AND CARTOONISTS CAPITALIZE ON TMI ACCIDENT HUMOR

HE'S GROWN A FOOT SINCE I SAW HIM LAST....

"When Maria said we could buy a condo on the beach for under $30,000, I said 'Where, 3-Mile Island?'"

Lead-in to Miami Beach Real Estate Developer Ad

CHAPTER V

LEGAL PURSUITS

"Lawsuit mania—a continual craving to go
to law against others while considering
themselves the injured party"

Cesare Lombroso Italian Criminologist

There was possibly no more frustrating, time-
consuming and expensive issue for GPU in the years
following the accident than the pursuit and defense
of legal actions stemming from that March '79
event.

While it was a foregone conclusion that extensive
legal claims would follow any accident of the
magnitude and public concern engendered by TMI,
they began almost before the reactor had cooled
down. They promise to continue after the plant has

been cleaned up to its safe, long-term storage condition.

Investigations of the accident and its root causes showed a number of parties contributed to the nation's first major nuclear accident. Investigative reports brought out facts that called for GPU to take legal action against the nuclear regulators, as well as the designer of the TMI plant. A group of shareholders felt the company had failed to disclose the magnitude of the financial risk associated with a nuclear accident and brought suit. Pennsylvania residents sued to recover business losses associated with their displacement during the accident period. Personal injury claims abounded, covering a multitude of alleged afflictions from nervous disorders to possible radiation induced illnesses.

Cleanup Costs—the Babcock and Wilcox Case

A primary issue for the company was the enormous cleanup cost. From the outset, GPU's position was that the burden had to be shared by all those involved in the nuclear venture. GPU defended this position throughout its legal battles, just as it did in its regulatory and financial affairs. On the regulatory front, GPU argued that the state and federal government, as well as the utility industry, should contribute to cleanup. And the Thornburgh plan did create a formula for just that.

On the corporate legal front, GPU argued that Babcock and Wilcox, as the engineering firm that supplied the nuclear steam plant and the operator's training, deserved to share at least a portion of the cleanup costs. In March 1980, GPU filed suit against B & W. for $500 million. However, GPU had already incurred damages in excess of that amount. By December 1981, GPU amended its complaint against B & W, increasing the damage claim to $4 billion.

GPU reached a settlement with Babcock & Wilcox in January of 1983. The terms of that agreement called for B & W to provide rebates of in the form of services and equipment purchases over the next ten to fifteen years. The proceeds from these rebates are being applied to the TMI-2 cleanup.

One aspect of the GPU/B & W lawsuit activity continued to plague the company for many months after the settlement. In its deliberations over granting approval to restart Unit 1 at TMI, the NRC pored over the documents and proceedings of the B & W suit. They contended there had been possible withholding of information and other alleged discrepancies related to the accident. This seriously slowed the NRC granting GPU the go-ahead on Unit 1. Unfortunately, these considerations were given substantial attention by the anti-nuclear community and the media before they were resolved and restart approval was given.

The NRC Suit

In keeping with its contention that the cleanup costs must be shared because the accident responsibilities were shared, GPU brought suit against the Nuclear Regulatory Commission. The federal agency responsible for oversight of the nuclear power industry, GPU argued, had been negligent in its duties. GPU alleged that the NRC violated both its statutory and regulatory duties to warn nuclear plant licensees of defects in equipment, analyses, or training procedures. In addition, GPU claimed that the NRC failed to inform the company of a similar nuclear plant incident that occurred before TMI—a warning which GPU believed could have prevented the TMI-2 accident. In fact, both the Kemeny and the Rogovin accident investigation reports tended to substantiate GPU's case against the NRC on this point.

As required by federal law, GPU first filed a $4 billion damages claim with the NRC in December of 1980. GPU claimed that the NRC's negligence contributed to the accident. But the agency was slow to react. The NRC waited until June 8, 1981, the deadline date, to respond with a denial of GPU's initial claim. Six months later, GPU filed its complaint against the NRC in federal court.

But, in September of 1984 the Third U.S. Circuit Court of Appeals turned down GPU's suit, contending that the NRC's actions before the TMI-2 accident fell within U.S. laws that protect the government from liability.

Re-start—the State Regulators and the Courts

Only a few months after the accident, the New Jersey Board of Public Utilities (NJBPU) removed the operating and capital costs of TMI-2 from the rate base of Jersey Central Power and Light. The Pennsylvania Public Utilities Commission acted similarly, removing that unit from the rate base of both Met-Ed and Penelec. The GPU subsidiaries filed complaints with the state regulatory authorities opposing these decisions.

After Unit 1, the undamaged unit, was ordered shut down by the NRC in July of 1979, the NJBPU removed TMI-1 from JCP&L's rate base in April 1980. Again acting in concert, the 'PaPUC removed TMI-1 from the rate base of both Met-ed and Penelec. These actions of the NRC and the state regulatory authorities regarding the undamaged unit seriously threatened the survival of the GPU subsidiaries.

In the ensuing years, GPU strongly contested these moves, but to little avail. In April 1984, the U.S. Supreme Court refused to hear JCP&L's appeal of a lower court's decision affirming the rate order removing both TMI units from the rate base. Although the issue of TMI-1's exclusion from the rate base was important, the company's chief objective had to be the expeditious restart of the undamaged unit. Quite simply, the position of the state regulators was that they would not return Unit 1 to the rate base until the NRC gave its approval to restart that unit.

Grand Jury Indictment of Metropolitan Edison

Probably the most devastating post-accident legal event to strike the company was an eleven count, federal indictment delivered up against Met-Ed in November of 1983.[1] These charges had been the subject of grand jury investigations since 1980 and involved the falsification of water inventory (leak rate) test data at TMI-2 prior to the accident.

---

[1] Indictment and accompanying News Release -issued by U.S. District Court for the Middle District of Pennsylvania; USA vs Metropolitan Edison Co., Nov 7, 1983

In February of 1984, settlement of these charges was reached. Under that settlement agreement, the company pleaded guilty to the charge of using an inadequate water inventory test procedure. The government dismissed four counts, including the only one that charged a violation of the criminal code. The company pleaded no contest to the remaining charges. Fines totaling $45,000 were levied against Met-Ed and the company agreed to fund a one million dollar program for emergency planning to directly benefit the area around TMI.[2]

These charges and the settlement were a painful admission, despite the prosecutor's statement that no officers or directors of GPU Nuclear or Met-Ed were involved.[3] Once the settlement was effected, the company could and did proceed with its own investigations. Those employees involved were disciplined and removed from reactor operation positions. Internal controls were imposed and safeguards strengthened to help insure that the type of behavior that resulted in the indictment would not recur. The nature of these events, the lengthy period before their resolution and the

---

[2] Plea Agreement, USA vs. Metropolitan Edison Co; U.S. District Court for the Middle District of Pennsylvania February, 1984
[3] Statement of Facts Submitted by the United States, USA vs. Metropolitan Edison Co, Criminal No.83-00188

extensive media coverage of the events slowed GPU's efforts to move ahead with cleanup operations and the program to obtain approval for the restart of TMI-1. This was a low point in the turnaround battle.

The Exam Cheating Incident

The all-important legal battle to restart TMI-1 was to prove longer and more arduous than anyone envisioned at the outset. Restart hearings began in 1980. New obstacles to a speedy resolution seemed to appear in rapid succession. One obstacle was the allegation of cheating on the TMI operators' licensing exam administered by the NRC during April 1981. An investigation followed that resulted in the resignation of two employees. The NRC then ordered all TMI-1 operators to retake the federal exams—a laborious and morale crippling task. GPU protested the action vigorously but to no avail. Support for restart was made more difficult by the substantial attention given to these events in the local press.

After nine months of hearings, the Atomic Safety and Licensing Board of the NRC did conclude that GPU had the managerial capability to operate TMI-1 without endangering the public. However, delays on other decisions relating to design modifications and emergency preparedness procedures, as well as further discussion of the cheating incident, continued to prevent the NRC from rendering a decision on restart. The debate eventually moved from the federal regulatory bodies to the judicial system

Psychological Stress Ruling

Hearings on the restart of TMI-1 were reopened in November, 1981, and the ASLB gave conditional approval for TMI-1 to begin low-power testing that December. But the new year brought a new obstacle. An anti-nuclear activist group, People against Nuclear Energy (PANE) contended that under the provisions of the National Environmental Protection Act (NEPA), the issue of psychological stress must be considered before the restart of TMI-1 could be approved.

The group took its case to court, and in January, 1982, the U.S. Court of Appeals for the District of Columbia ruled that TMI-I could not be restarted until the NRC assessed the possible psychological impact on local residents.

GPU petitioned the U.S. Supreme Court to review this ruling, and the court granted the petition in

November, 1982. Five months later, in April, 1983, the U.S. Supreme Court ruled unanimously that under the provisions of NEPA the NRC did not have to consider psychological stress before the restart of TMI-1 could be approved. The potential effect of this decision had implications well beyond the TMI case and even beyond the nuclear power issue.

With the psychological stress issue resolved, optimism increased for the restart of TMI-1. But several technical issues related to mechanical operations as well as some lingering competency concerns remained. Among these were allegations of procedural violations and unsafe practices during the clean-up of TMI-2.

Return to Power

Two years after the Supreme Court Decision on stress the NRC finally reached a decision, voting 4 to 1 to lift the shutdown order on TMI-1. But then a federal district court in Philadelphia issued a stay order, based on petitions filed by several anti-nuclear activists as well as Pennsylvania Governor Dick Thornburgh.

The U.S. Supreme Court was once again called upon, voting in early October, 1985 to lift the lower court's stay. The 8 to 1 vote made it a decisive victory after six long years!.

At 4:02 a.m. on October 9th, TMI Unit 1 went back on the grid supplying electricity at 15% of power, generating about 50 megawatts. In January of 1986, the unit completed its startup period reaching 100% power, providing 880 megawatts of electricity— enough to supply more than 500,000 homes.

Shareholder Suits

The shareholder lawsuits alleged the liability of GPU's corporate directors for failing to disclose in published reports, issued prior to the accident, that the possible outcome of a nuclear accident could be severe financial hardship.

In fact, GPU's position on shared liability spoke directly to the shareholder's complaint: if all costs of the accident had to be absorbed by the company alone, then it was ultimately the shareholders who were incurring extraordinary risks. It is a problem engendered by the hybrid nature of a public utility company owned by shareholders and regulated by state government. Regulatory commissions and the financial marketplace have as yet not fully come to grips with this problem. As GPU's General Counsel Jim Liberman put it:

> "Nuclear power is really a joint venture between the people, the government, and the utility industry—until there was an accident, at which point it became not a joint venture but the entire responsibility, and therefore the entire burden of the utility industry. It has never again been a joint venture."

The class action litigation brought by shareholders was settled in September, 1983, when a federal judge approved a settlement reached by GPU earlier that year. The terms included a payment of $6 million to an escrow account for the plaintiff class and issuance of 1.6 million shares of common GPU stock for partial payment of the settlement.

Personal Injury Suits

Finally, individual and class action lawsuits alleged personal economic, health or property damages. In February 1981, a federal Judge gave approval to a $25 million settlement for the individual and class action suits, with $5 million set aside for the establishment of a public health fund. The administration and use of that fund has been the source of numerous disputes.

Early in the post-accident legal proceedings, the courts had decided that personal injury suits would not be pursued in class actions. In 1985, 292 personal injury claims were settled by the company's insurers for an aggregate of $14.3 million. Approximately 2100 personal injury claims most filed after these settlements, remain pending. In an effort to reach a resolution of this large number of claims, twelve "test cases" from among the 2000 have been selected, six by the plaintiffs and six by the defendants. The verdict in these 12 cases will determine the outcome for the majority of the pending cases. As of mid-1988, some of the plaintiff's cases have been dismissed by the Pennsylvania state Superior Court because the complaints were not filed within the two year statute of limitations. No trial date for the 12 test cases has been set.

For an accident that caused no casualties, and according to reports of government agencies did not constitute a significant public health or safety hazard, the amount of litigation has been enormous. Perhaps it is now drawing to a close.

CHAPTER VI

REBUILDING COMMUNICATIONS

"Sensationalistic reporting of all the
claims and stories following the
accident, without bothering to check the
source or investigate the details, was
clearly damaging to our recovery."

W.G. Kuhns Chairman, GPU

Unprepared For Crisis

On the day of the accident, GPU's crisis
communications program was adequate by the
standards of the day—both by NRC and industry
standards, as well as by the standards of the
business community at large. But it was by no means
adequate to handle the accident.

Just as TMI was a "first" at the technical and regulatory levels, it was a first in the area of crisis communications. While the sheer size and antagonism of the media would have swamped most communications departments, it was certainly beyond the capabilities of Met-Ed, the operator of TMI. The utility had six professionals in its communications department, three of whom were assigned to media affairs. Only one person was actually stationed at the TMI site, primarily to conduct plant tours. As Bill Kuhns bluntly put it:

> "We were not prepared for such an event
> and were simply overwhelmed by it."

Prior to the accident, literature provided by the company to the public on nuclear power was more promotional than technical, since the industry had never been eager to share much technical detail with the press and public. Actually, the press was disinclined to want to hear much about the technical operations of a nuclear plant. For the most part, reporters were also not trained to understand the subject in any depth.

A research analysis of press coverage of the TMI accident done five years later at Lehigh University describes the pre-accident relationship between company communications and local press coverage:

"TMI coverage before the accident included such topics as antinuclear protests, plant evaluations, licensing hearings, safety and plant events and security problems...A number of articles related 'good news' about TMI including the dedication of TMI-2, a good rating for TMI-1 from the NRC and the beginning of commercial operation of TMI-2.

Pointedly missing was detailed coverage of problems occurring at the nuclear plant, despite the fact that Met-Ed issued weekly press releases detailing the status of each reactor and what problems were reported that week. They were ignored, partially because they were written by engineers who used technical jargon and partially because the newspapers weren't paying a great deal of attention to the TMI story."[1]

In short, neither the company nor the press was well prepared to deal with this accident. To that fundamental deficiency one must add the very high emotional content of the event.

Often during the company recovery period, the question was raised as to why the media reacted so

---

[1] Sharon M. Friedman, "Local Coverage of Three Mile Island During 1981-82," pg. 7. Presented Association for Education in Journalism and Mass Communication Annual Convention, University of Florida, 1984. (Research was funded by a grant from the Energy Research Center and the Office of Research, Lehigh University).

strongly to the TMI accident. Bill Murray, GPU VP for Communications commented on this point:

> "For several years the nuclear industry was arrogant toward reporters probing questions about the safety of nuclear power. Whenever the press raised a concern on safety, they were more or less told that they wouldn't understand the required technical answer—just take our word for it. The press did not like that. They were waiting for something that would puncture that balloon. Once nuclear power was shown to have a fault, a weakness, they were ready to pounce."

The communications situation that GPU faced in the aftermath of the accident was overwhelming. Hundreds of reporters were sent to TMI to cover what was being called "the biggest story of the century." When they got there, there was little information available for a story. The media wanted volumes of detail for extensive stories of this first nuclear accident.

What was available failed to satisfy the hunger of the press. A two page release issued by Met-Ed announced the occurrence. At the time, it was looked upon as barely scratching the surface. Later, the NRC said it was the most reliable and accurate information that came out in the first three or four days into the accident. The press,

however, was looking for much more. In their
frustration, reporters stood at the plant gate and
grabbed workers as they came out. In some cases
they even went to workers' homes to get interviews.

Today, the TMI accident is studied as an example of
crisis communications. It is easy to look back and
say that if the company had only had literature
prepared in advance, written in laymen's language,
describing the plant and its operation, it could
have filled this information gap and saved itself a
lot of grief. Yet, it's impossible to say under the
circumstances whether any amount of available
information would really have been enough.

Communicating a Dearth of Information

Technical facts on what was actually happening
inside the reactor dribbled out to not only the
communication staff but to the engineers involved
in trying to cool down the plant.

Initial emphasis was properly placed on shutting
down the reactor, not figuring out the cause of the
accident.

Unfortunately, the Met-Ed and GPU staff really
didn't know exactly what was happening, and in some
cases were reluctant to pass on the little

information they did have. Once the first two press releases went out, the information flow became a trickle. The company was then viewed as withholding information. Bill Murray recalls those days:

> "The press never understood the problem we had in not being able to get them sufficient information. We didn't know the cause of the accident or what its outcome would be. But the reporters were unable to believe that. It was after all a post-Watergate press eager to find a cover up."

Regulatory Problems

As quickly as possible, Met-Ed set up physical facilities at the site in an attempt to handle the thousands of media inquiries that were pouring in. Press conferences, that turned into shouting matches, were held. Telephone lines were expanded, staff from all the GPU companies were sent to the site and a large media center was set up for reporters. All of the measures proved insufficient. The NRC compounded the company's problems with the media by the very nature of its regulatory role.

NRC press conferences were held several hours before or after company meetings with the media, sometimes producing information and off-site radiation readings that differed from GPU's. This was understandable given that the situation changed quickly and the technology was complex. But it made the company appear unreliable at best and duplicitous at worst.

GPU asked the NRC to hold joint press conferences so that more thorough explanations could be given and any conflicting facts explained. But the NRC refused, saying it was important that they maintain their position as regulator entirely separate from GPU, the regulatee. Media antagonism grew. Finally GPU felt compelled to stop holding separate press conferences, turning this function completely over to the NRC. The credibility of Met-Ed and GPU with the media was disappearing rapidly. It would be many months before it began to be regained.

Emotional Words

The lack of enough hard data, the inadequacy of staff, the conflicting reports and the very nature of the subject all contributed to a heightened emotional reaction to the situation. Some of the media, however, chose to move from emotionalism to sensationalism.

Some examples suffice to convey the tone. Forty-seven reporters from one paper alone covered the story and in April of 1979 ran the following headlines:

"Solace in prayer on a day of terror"

"A life or death mission breaks a worker's routine"

"A child's fears:" The big ball killed everybody'

A New York reporter ran a story in which he described seeing radiation running down the sides of the cooling towers.

But even newspapers that did not sensationalize tended to emphasize the negative, with little ever said on the positive side of the story. For instance, there was little coverage of the company's extraordinary success in quickly bringing a score of the nation's nuclear power experts to TMI to discuss the accident and advise on placing the reactor in a safe shutdown mode. On the other hand, the hydrogen bubble scare received enormous coverage. One week after that story was defused, a local TV station, during a re-run of a soap opera, neglected to remove a banner news line from the previous week that ran across the screen repeating the threat of a hydrogen explosion. Calls bombarded the company as the public was unnecessarily panicked again.

Lingering Consequences

After the initial shock of the accident, the first communications task was to build up the staff at the Three Mile Island site. GPU Nuclear increased the communications group to about 35 people. Experienced reporters were brought on board to interface with the media and start the long, slow process of regaining credibility.

The TMI accident changed the way all nuclear facilities must communicate. Media relations managers for other nuclear plants looked at the TMI experience at that time and were heard to say, "There but for the grace of God go I." TMI heightened both the public's and the media's awareness of nuclear power, and also their mistrust. Given this enhanced awareness, combined with the added difficulty that a generally uneducated press posed during the accident, GPU embarked on a policy and a program to make the press and the public as fully informed and knowledgeable on nuclear matters as possible.

No event or activity at TMI or by GPU was too small or too insignificant not to be completely and candidly reported to the world at the earliest possible moment and in words that the public could understand. On complex stories, special media education sessions were held. Plant tours for the press and public were greatly expanded. Any group, pro or con, was given the opportunity to visit the site and participate in briefings and discussions. The media center was refurbished with expanded exhibits on the accident and on nuclear power in general. A new film was produced for public viewing. Speakers bureaus, town meetings, and community briefing sessions were established.

One very effective communication method used was a quarterly forum where GPU representatives answered questions from the public and the NRC. The company panel was composed of people who offered a balance between technical and non-technical backgrounds. As Bill Gifford, VP of Communications at the TMI site for much of the recovery period, expressed it:

> "Organizing and obtaining community support is probably the most effective approach that a company can employ on complex and controversial issues. But it must be completely candid and sincere."

This Is A Drill

The TMI accident taught GPU and the world the need for greatly increased training and practice for coping with communication catastrophes. Accidents since TMI have vividly demonstrated this need. Witness more recent oil spills, food contamination scares, toxic waste dumps, high rise fires, as well as other nuclear plant incidents. How often do you still read in post-accident analyses of such events of a "failure of communications?"

GPU embarked on an intense program of realistic drills simulating various levels of possible nuclear crises. Planned as well as unplanned exercises were repeatedly held and strictly critiqued to ensure readiness. This simulation method of crisis management training has now been picked up and used by all forms of business, not just the nuclear industry.

Krypton Venting

One major GPU press and public relations event stands out in the post-TMI accident period because of the great fanfare surrounding it and because its outcome represented one of the first positive moves in the company's climb back. This was the venting of krypton gas from the damaged reactor building.

Krypton gas, produced during the accident, had to be released from the TMI building before any major cleanup operations inside could be accomplished by re-entry teams. A plan for a safe and controlled venting to the atmosphere was proposed, and the company spent many months obtaining approval for this operation. The press had a field day with the krypton venting proposal. It became a cause celebre for anti-nuclear groups, local and national. Press coverage was particularly negative. Finally, in June of 1979, it was approved and the venting was carried out safely, with insignificant impact on the environment and with no harm to public health. In fact, on the day of the venting, hundreds of company employees and their families came to the visitor center just across the river from the plant for a picnic with the press. A turning point of sorts in positive company communications had been reached.

Restart Communications

The effort to obtain approval for the restart of Unit, 1 at TMI became another rallying point for the anti-nuclear community's communications. The company similarly placed a high priority on getting word to the public on its position on the restart issue. Concern over the possible restart of a nuclear plant in the TMI area raised all of the old issues and several new ones. There were technical points regarding possible embrittlement of the reactor vessel and leaks in steam generator tubes. Many legal moves were made to block restart. A unique issue was whether "psychological stress" should be a consideration—a question that ended up in the U.S. Supreme Court, as discussed in an earlier chapter. All of these issues provided considerable fuel for stories, editorials and cartoons in the press. The majority of these had a decided leaning against the company and restart. It was an uphill fight.

An almost classic example was the case of the Aamodts. This local area couple, with no scientific credentials in the field, produced a study claiming a dramatic increase in cancers among the inhabitants of the TMI area. The U.S. Center for Disease Control's analysis concluded that the Aamondts' methods were unsystematic and subject to interviewer bias. The Pennsylvania Health Department also said that the Aamondt study was "flawed" and the Department's own epidemiological studies did not provide evidence of increased cancer risks to residents near the TMI nuclear plant. However, the initial NRC recognition of the Aamodts' survey, and the widespread media attention, gave this incredibly bad study a level of importance it did not deserve.

Company Ads

One series of articles during the restart period by a Philadelphia paper contained so many blatant errors and inflammatory comments that the company felt compelled to mount a paid advertising campaign to inform the public of the facts and counteract the flood of misinformation. Full page newspaper ads were taken out specifically calling out the errors contained in the news stories. These ads were later followed by a series detailing the company's moves to ensure public health and safety. Newspaper ads were accompanied by a program of short TV spots that ran in the local area. Although decried by the anti-nuclear activists as a PR blitz, later polls indicated the ads had been helpful in favorably moving public opinion and bringing GPU one more step back in its recovery.

Employee Power

During this communications credibility recovery period, it was demonstrated again and again that a key factor was the role played by GPU's employees at all levels. Success or failure was heavily dependent upon keeping all employees well-informed and highly motivated.

Daily and weekly publications, call-in phone lines, multiple meetings with management, video program updates, tours including family members—all were designed to keep employees up to the minute on every relevant event. These efforts resulted in the employees becoming the best ambassadors to the public that GPU could have.

Climbing Back

The initial loss of credibility with the press and the public remained a nagging burden for GPU during a large portion of the recovery years. The company's reputation and integrity was challenged many times. Technical and organizational problems with the difficult cleanup operation, the exam cheating scandal, the lawsuits, the indictment of Met-Ed over leak rate data, the conduct of emergency drills and a host of other "unusual events," as they are labeled by the NRC, combined to stagger GPU just as it would seem to be getting up off its knees.

Measuring Success

By and large, GPU and its nuclear operations have today regained the confidence and trust of most of the press and public. Nevertheless, success is not complete and probably never will be. Electric utilities and nuclear power cannot expect to have completely smooth sailing. There remains both a local hard core of anti-nuclear activists and a general anxiety world-wide that must be recognized. In the TMI area, these can be counted on to provide the press with an anti-GPU or anti-TMI statement on almost any occasion.

But accident anniversaries are now bringing out only a handful instead of hundreds of demonstrators. Newspaper editorials contain praise for GPU's recovery and the concern it has shown for public health and safety. Politicians, pondering the successful operation of TMI-1 since its return to service, are far less likely to go for the nuclear jugular, particularly in view of rising concerns over acid rain, the greenhouse effect, and brownouts.

All is not completely serene on the TMI scene. Some concern remains over the disposal of stored accident water and the plans for long-term, monitored storage of the TMI-2 plant. However, these issues are being discussed and debated in a far more reasoned atmosphere. For a change, such issues are now generating "more light than heat." A strong communications rebuilding program has proven its worth.

CHAPTER VII

CHERNOBYL AND TMI:

MORE DIFFERENCES THAN SIMILARITIES

"..In comparison with Chernobyl, Three
Mile Island was just a trifling
accident."[1]

J. Jovanovich Professor of Physics
University of Manitoba

Although there is general agreement with the above
quoted opinion, so much has been written, and
probably will continue to be written, about these
two nuclear events that it seems proper that the
story of GPU's recovery should comment on them.

---

[1] San Diego Union, April 26, 1988

Since the April 1986 nuclear disaster at Chernobyl, there has been a growing body of literature linking, comparing and contrasting that event and TMI. And there are some similarities—and most assuredly many differences.

It is agreed today that, among other causes, both of these accidents were the result of a combination of human error and equipment failure.

The Chernobyl accident was reminiscent of a similar problem at TMI—human error resulting from "operator mindset." The most serious mindset was a preoccupation with the safety provided by equipment, and the consequent downplaying of the importance of the human element in safe nuclear plant operation. Ten years ago, as the Rogovin and Kemeny reports on TMI made clear, this was a problem throughout the U.S. nuclear power industry. Fortunately, as a result of the TMI studies, the U.S. learned of the "mindset" deficiency and worked to correct it without the casualties suffered in the Russian accident.

The most significant difference between TMI and Chernobyl lies in the basic design of the two plants. The designs are quite different, with the result that the Russian reactor places the workers and the general public at much greater risk. For example, Chernobyl did not have the complete containment that prevented a major escape of fission products from TMI. Also, the Chernobyl plant design has rather unstable operating characteristics as the temperature in the reactor rises, whereas in U.S. commercial nuclear plants the reactivity goes down as the temperature goes up. In terms of design, the Russian reactors are simply not as safe as the U.S. plants.

One of the early comparisons with TMI to appear in the media after Chernobyl concerned the comparative handling of public information at the time of the accidents. Anti-nuclear writers quickly went to press with stories telling how the U.S. had been just as slow in its information release at TMI as the Russians were at Chernobyl. Even Gorbachev, in his initial address to the Russian people on the accident, made that comparison. After all, these stories went, wasn't the seriousness of the TMI accident withheld from the world in the same manner as Chernobyl was not acknowledged by the Russians until radiation was picked up by foreign monitors?

Fortunately, more objective writers shortly prevailed and pointed out, as did William Safire in the New York Times,[2] that Americans opened "floodgates of information" on TMI to the world, compared to the Soviets. And extensive investigations have shown the initial shortage of information at TMI was due to the lack of precise knowledge as to exactly what was going on inside the reactor, not an effort to withhold data.

But information-sharing has, in fact, been one positive outcome of the Chernobyl accident: there is now a growing program for an effective international network of information-sharing on even the most minor nuclear events.

A radical difference between the two accidents may be seen in the time taken before the restart of the undamaged reactor units located on the accident sites in each of the two countries. The undamaged reactors on the Chernobyl site were back running within months after that accident. The undamaged Unit 1 at TMI fought a six and a half year battle, primarily political, to reach that point. One could speculate how long the approval to restart Unit 1 at TMI might have been even further delayed if that unit had not already been given the go-ahead before the Chernobyl catastrophe.

---

[2] Reader's Digest, August 1986-Condensed from the N.Y. Times

A startling comparison is the differing radiation levels reported in Middletown, Pennsylvania as a result of these two nuclear accidents. Robert Bleiberg wrote in <u>Barrons</u> on July 20, 1987:

> "...last year's Chernobyl disaster in the Soviet Union spread by the winds throughout the Middletown, Pennsylvania area, levels of radioactive iodine (perhaps the most widely feared nuclear element) three times higher than did the TMI-2 mishap.
>
> In both cases, the iodine levels were low and posed no hazard to health."

In other words, the Chernobyl accident spread moreradioactive iodine in Middletown Pennsylvania than did the neighboring accident at TMI!

The impact of both TMI and Chernobyl on the near-term future of nuclear power was undeniably very negative. The direct effect of Chernobyl on TMI itself, and on GPU's comeback, was minimal, however, with the exception of the media comparisons.

CHAPTER VIII

FACING THE FUTURE

"In terms of turning the company around, they're not dealing with problems anymore; they're trying to position themselves for the future."

Linda S. Caldwell Analyst,
Duff and Phelps

This analyst's comment, which accompanied Duff and Phelps upgrading of GPU securities in early 1988, summarized the position of General Public Utilities nine years after the Three Mile Island accident.

The TMI accident era was drawing to a close. Unit 1 at TMI, generating at full power after the long approval fight, had just completed its most successful operating run ever. The operation at Unit 2 could see successful placement of that plant in a long term, monitored storage mode in 1990. The costs of that difficult operation looked as if they would come in at about the level projected seven years earlier.

The company's Oyster Creek nuclear plant, approaching its 20th year of operation, had completed a record 160 day run at full power.

Earnings were well above the previous year's levels. The dividend had been tripled in 1988 and was back at pre-accident levels. GPU common stock waslling at all time highs. Buy recommendations were common on Wall Street. Private placement of operating company bond issues had been made in 1985 and '86, and the first public bond offerings since the accident were made in early 1988. Customers rates were among the lowest in New Jersey and Pennsylvania.

There was no doubt that General Public Utilities was turning its primary attention to the future, well-positioned to take on the opportunities and challenges facing the utility industry at the end of the decade of the '80's.

GPU, as well as the entire utility industry, learned much from the accident at TMI. Many of the lessons were positive. The company was strengthened organizationally and financially by going through the ordeal of near-bankruptcy and recovery. As the chapter on TMI drew to a close, GPU could look back with pride on its accomplishments and confidently focus on the future.

The power supply strategy of GPU in the upcoming years adds up to four major programs, as retiring President Herman Dieckamp described it at the May 1988 shareholders' meeting:

> "First, extend the lives of existing generating stations; second, extend power purchase contracts; third, encourage cogeneration and other forms of independent power production; and finally, minimize the need for new plant construction."

Assessing GPU's financial future, Chairman Bill Kuhns told GPU shareholders early in 1989 that:

> "GPU is making considerable progress toward achieving its four financial goals of controlling expenditures, retraining capital expansion, maintaining solid 'A' bond ratings and managing the system's equity."

As the decade was drawing to a close, the GPU system had already made several moves to take advantage of the changing world facing its industry. A new subsidiary (General Portfolios Corporation) was formed as an umbrella corporation for non-utility and investment business. Economic development efforts in the GPU service territory were being expanded. Cogeneration, conservation and load management continued to receive prime attention, as their contribution to holding down future capital costs for new generating plant construction was recognized. Overall, GPU was working with regulators in a joint effort to shape electric power policy and build public confidence in a new partnership of consumers, regulators, entrepreneurs and utilities for the `90's.

In the coming years. GPU can expect to see more changes in the utility industry as an era of increasing deregulation moves ahead. Careful adjustments to these changes will be required by an ever-alert management. Sources of non-owned generation will become available. Opportunities for investment in businesses other than the regulated utility business will arise. Competition will increase. Cost-control and efficiency of operation will continue to be essential.

William Murray

Nuclear Power's Outlook

As GPU assesses its future, both nuclear and non-nuclear, the future of nuclear power overall bears examination in light of the TMI experience.

As has often been pointed out, no U.S. utility has ordered a nuclear plant since the TMI accident. It is unlikely that there will be any such orders for possibly 10 to 15 years. Today, nuclear power seems caught between the twin shoals of a lack of public acceptance and unacceptable costs. Solutions lie in the development of simpler, standardized and more forgiving plant designs, greater construction pre-fabrication, and a more streamlined licensing process. Work is proceeding in all of these areas but it will take time.

The Risk Factor

Compared with many other industrial risks to the population, nuclear power has a remarkable safety record. Unfortunately, it is generally perceived as highly unsafe. A 1986 Wall Street Journal editorial cited the nuclear power industry as the leading victim of what it called "science and demagoguery" in this country. The editorial stated:

> "The public regards itself as poorly trained to judge (scientific) issues and

receives   little   information—from
researchers, the press or protesters—
adequately   describing   the   relevant
science. Public interest groups exploit
this vacuum, and their poorly supported
accusations, appearing in stories under
the over-used rubric "critics charge",
cast   doubt   over   technology   and
science."[1]

In the wake of the TMI and Chernobyl accidents,
psychologists and sociologists were called on to
explain why people had responded to the risk of
nuclear plant accidents with such vehemence. While
causative factors are often hard to find, some
patterns seem clear.

First, people tend to overestimate the probability
of unfamiliar, catastrophic and well-publicized
events and underestimate the probability of
familiar events that claim one victim at a time
(e.g.: commercial air flight vs. automobile
travel). In addition, people respond most
negatively when the precise degree of risk is
unknown and the consequences particularly dreaded
(e.g.: dioxin contamination). And lastly, voluntary
risk is accepted much more cheerfully than
involuntary risk (e.g.: sky diving or cigarette
smoking vs. air pollution).

---

[1] Wall Street Journal, July 31, 1986

Not surprisingly, there is a direct correlation between the degree of risk perceived and the frequency with which a potential risk is mentioned in news reports. The way in which risk is stated is also a factor. Consider the difference between saying that one's personal risk of disease is doubled versus saying that the risk of the disease in the population has risen from one in one million to two in one million. Finally, and again not surprisingly, there is research showing that beliefs about risk are slow to change and show extraordinary persistence in the face of contrary evidence.

The task of educating the public and obtaining public confidence will be a long and arduous one. The challenge to the nuclear utilities will be to demonstrate safety and efficiency through outstanding plant performance year after year.

The Economic Hurdle

Possibly even more difficult to overcome than public antipathy will be the current poor economics of building nuclear plants. In the years that followed the accident at TMI, nuclear plants still under construction showed higher and higher costs for a variety of reasons. Cancellation of plants still on the drawing boards soared. Nearly completed plants such as Shoreham and Seabrook were caught up in the anti-nuclear flood tide.

In June of 1988, U.S. Energy Secretary Harrington said that electric utilities were "on notice" of licensing obstacles from foes who "took cheap nuclear power and battered it into becoming expensive in a self-fulfilling prophecy."[2]

According to the U.S. Council for Energy Awareness (USCEA), almost 80 percent of Americans today believe nuclear energy will be important in meeting our electricity needs in the years ahead, and 53 percent say it will be very important. Yet, considering all of these factors, it appears that it will probably take some major event down the line to bring about the reordering of nuclear power plants by the utility industry. Rising global concerns over the "Greenhouse Effect" during the summer of 1988 were beginning to show what such an event might be. As Newsweek headlined it: "Not so Bad After All?—Nuclear Power Revisited".[3]

A step-up in the acid rain battle or a significant shortfall in the estimated energy supply from cogeneration and other alternate energy sources could also push the return to nuclear power.

---

[2] San Diego Union, June 14, 1988
[3] Newsweek, July 25, 1988

The GPU system reached new all-time power peaks several times during the summer of '88. Utilities throughout the Northeast found themselves running at peak load levels they had previously forecasted reaching only by the mid 1990s.

The significance of this for the future of nuclear power was noted by President Reagan in August of 1988 on the occasion of his signing an extension of the nuclear liability law when he said:

> "I sign this legislation in the midst of a summer that has brought record temperatures to much of our country. As a consequence, many of our utilities find themselves near the limits of their power generating capacity. The implication of this situation is clear: Our nation must move forward into a new era of safe, economical, and clean nuclear power."[4]

A few weeks earlier Chairman Bill Kuhns of GPU summed up the future for nuclear power as follows:

> "Many energy experts agree that we need nuclear power. The non-nuclear sources are finite and heavily laced with political considerations. The "Greenhouse Effect" will be the subject of increasing attention and will underline the importance of non-fossil generation. As that scene develops, the

---

[4] Wall Street Journal (AP), August 22, 1988

availability of a new, simpler, standardized, more forgiving design of smaller, more prefabricated and prelicensable nuclear plants will provide an attractive option."

EPILOGUE

Since this book was written, several events have occurred that add significance to this narrative.

In April 1990, the last of the nuclear fuel from Unit 2 was shipped to Idaho. In September,1990 researchers at Columbia University reported finding no excess cancer from radiation releases from the accident. A National Cancer Institute study reaches a similar conclusion.

In June 1996, a Federal judge dismisses all 2100 lawsuits claiming injury from the accident citing slim proof.

Yet, U.S. utilities have not ordered a single new nuclear power plant since the accident. This is primarily because they did not wish to take the big financial risk of building such a large facility.

However, things are changing. Nuclear plant designers have come up with new, smaller plant designs incorporating many of the safety and reliability changes adopted by the electric utility community since the TMI accident.

In recent years, utilities found they could meet their growing needs with smaller, natural gas fired units. Now, we are beginning to realize that we will be running short of natural gas supplies for this purpose. Nuclear powered units of smaller, more efficient, safer designs are looking attractive.

The current administration in Washington recently introduced and the Senate endorsed a plan for the government to provide loan guarantees for the construction of a half dozen nuclear power plants. Sen Pete Domenici, the architect of the program said: "The time has come to quit playing around with energy, and say wherever we can we are going to produce more energy." Nuclear power has long been neglected and that has been "a giant mistake" according to Sen Domenici. Obviously, such a plan will generate much interest and comment from the pro and anti-nuclear groups in the country in the coming months. There are many lessons to be learned from the successes of General Public Utilities in handling the problems arising from the TMI accident. They will be applicable to future decisions.

The comments and conclusions of the key personnel involved in the accident and the recovery are of particular value because the events were fresh in their minds when the interviews were taken and the book was written.

Now, as we move into a serious period of considering adding to the nation's energy supply through nuclear power, it is important that we take advantage of the experience of the past as related in this document.

W.B.M., 2003